EXPLORING ST. CROIX

by

SHIRLEY IMSAND

and

RICHARD PHILIBOSIAN

Travelers Information Press
1385 Capinero Drive
Pasadena, California 91105
United States of America

For John A. Yntema

Library of Congress Catalog Card Number: 87-50208

International Standard Book Number: 0-9617976-0-6

TABLE OF CONTENTS

Cover photograph: Baobab Tree and ruins of sugar factory at Estate Butler Bay.

ACKNOWLEDGMENTS 6

HOW TO USE THIS BOOK 7

HISTORY 9

TOURS FOR CRUISE SHIP PASSENGERS 13
CENTERLINE TOUR 13
TOUR OF FREDERIKSTED 13
WEST SHORE TOUR 13
FOREST DRIVE 13
SCENIC DRIVE 13
BUCK ISLAND 14
EAST END AND SOUTH SHORE TOUR 14

CENTERLINE TOUR from CHRISTIANSTED to FREDERIKSTED
INCLUDING ROUTE 70 and PARTS OF ROUTES 75 and 753 17
Christiansted Tour 17
Hermon Hill Road and Questa Verde 22
Route 70 24
Route 682 25
Route 709 27
Route 663, Route 70 to Route 66 29
Route 707 30
Route 69, Route 70 to Route 66 34
Route 705 35
St. George Hill 38
Long Point Mangrove 39

SIDE TRIPS OFF CENTERLINE ROAD, ROUTE 70:
SIDE TRIP TO EXPERIMENTAL FOREST VIA ROUTE 708 43
Routes 811 and 81 43
SIDE TRIP ALONG FOUR-LANE HIGHWAY, ROUTE 66 46
Alucroix Channel Road 47
SIDE TRIP ALONG HESS OIL FENCE, ROUTE 68 52
SIDE TRIP ALONG AIRPORT ROAD, ROUTE 64 54
Road to Krause Lagoon 55
SIDE TRIP TO SANDY POINT 57
TOUR OF FREDERIKSTED
AND CONNECTING LINK TO ROUTE 63 60
Road from Frederiksted along the SW coast 62
Frederiksted Pier 64

WEST END ROAD, ROUTE 63 71
Route 702 to E 71
Estate Mount Washington Road 74
Hike up Butler Bay Gut 75
Hike to Ham Bluff Lighthouse 80

SIDE TRIPS OFF WEST END ROAD:
SIDE TRIP ALONG MAHOGANY ROAD, ROUTE 76, AND MIDLAND ROAD, ROUTE 72 81
Mahogany Road, Route 76 81
Route 763, Oxford Road 82
Route 765, Orangegrove Road 83
River Road, Route 69 86
Midland Road, Route 72 86
"To 70" 87
SIDE TRIP ALONG CREQUE DAM ROAD, ROUTE 58 88
Nicholas Road 92

NORTH SHORE TOUR INCLUDING PARTS OF ROUTES 75,80,69, AND 78 WEST 93
North Road, Route 75 93
Back Road to Experimental Forest 93
Little Princess Beach area 95
Route 74 95
Route 81 95
Route 751 to Salt River Bay 97
Route 79, Pepper Tree Road 98
Cruzan Motors horticultural plantings 99
North Shore Road, Route 80 101
Glynn Farms Road 102
Columbus' Landing 104
Estate Clairmont Road 107
River Road, Route 69 114
Scenic Drive West, Route 78 115
Furnel Ridge Hike 116
Roads of the Annaly Area 122

SIDE TRIPS OFF NORTH SHORE TOUR:
SIDE TRIP ON SCENIC DRIVE, ROUTE 78 127
Scenic Drive East, Route 78 127
Windwood Estates Loop 128
Canaan Ridge Road, Route 73 129
Scenic Drive West, Route 78 129
Vallee View Road 130
SIDE TRIP ON CANAAN RIDGE ROAD, ROUTE 73 131
SIDE TRIP ON RIVER ROAD, ROUTE 69 133

EAST END TOUR INCLUDING EAST END ROAD, ROUTE 82, SOUTHSHORE ROAD, ROUTE 60, AND SOUTHSIDE ROAD, ROUTE 62 137
EAST END ROAD, ROUTE 82 137
Road to Recovery Hill 137
Road to Altona Lagoon 139
Road to Darby Hill and Buccaneer Hotel 142
Tamarind Reef Road 145
Miss Bea Road 147
Road to Coakley Bay Saltpond 148
Road around the Reef Condominium to Mt. Tallin 152

East End Beaches 156

SOUTHSHORE ROAD, ROUTE 60 159
Grapetree Bay Hill Road to Romney Hill 159
Grapetree Bay area 160
Great Pond area 166
Route 624 167
SOUTHSIDE ROAD, ROUTE 62 167
Lowry Hill Road, Route 62 (northern leg to Route 82) 167
Maria Hill 167
Manchenil Bay Road 171
Route 83 173
French Pond and Canegarden Bay 173

SIDE TRIP OFF EAST END TOUR:
TIPPERARY ROAD, ROUTE 60 175
All-for-the-Better Road 175
Greatpond Hill Road 176

BUCK ISLAND NATIONAL MONUMENT 179

INDEX TO ORGANISMS AND SCIENTIFIC NAMES 181

GENERAL INDEX 191

Fig. 1. Sprat Hall Mill. What is the purpose of the long narrow opening? See p. 120, mile 13.5(6.0).

ACKNOWLEDGMENTS

We have spoken with many residents of St. Croix and we are grateful to them for their helpful ideas, their warmth, and their enthusiasm for this island. We are particularly indebted to the following people for many enjoyable and enlightening discussions about St. Croix: Anthony J. Ayer, William F. Cissel, Karen and Scott Eckert, Axel L. Frederiksen, William and Elizabeth Gladfelter, Margaret Hayes, the James Hurd family, Walter Knausenberger, Hans Lawaetz, John and Nancy Ogden, Earl Roebuck, G.A. Seaman, Fred W. Sladen, Anton Teytaud, Bruce E. Tilden, and alphabetically last but of paramount help, Margo and John Yntema.

Fig. 2. Teak trees in the Experimental Forest. The largest leaves grow on the younger trees.

HOW TO USE THIS BOOK

This book is a detailed guide for over 200 miles of roads and many miles of hiking. To help you choose which areas to drive, we have provided a brief description at the beginning of each main tour and each side trip off a main tour. You can also scan the text of any route, looking for the bold print which represents points of interest. Look through the general index to find items that interest you. Visitors who have only one or two days on St. Croix, such as cruise ship passengers or travelers making a short side trip from a neighboring island, should read the next section "Tours for Cruise Ship Passengers". There we recommend certain tours and their highlights, knowing that you may have to return to your ship or plane the same day you arrive.

The tours are organized so that you can start at the beginning or the end, whichever is more convenient depending on where you are. The first column of numbers has the odometer readings for the tour going in the direction listed at the top of the column; for example, E-W is east to west. If you are driving the tour in the opposite direction, use the odometer readings in the second column, starting with 0.0 at the "end" of the tour. Intervals between stops are given in pointed brackets, <>; information on connecting roads is given in square brackets, []. The text is written as if you were going in the direction listed for the first column. So if you are going in the opposite direction and following the second column, read the stop descriptions from the bottom of the page to the top, and turn the pages going toward the front of the book. All navigational directions in the text are for the principle (first) direction column. If doing a tour backwards (second column), one should use right turns where "left" are given and left turns where "right" is specified. We have also given compass directions wherever a turn is indicated. All compass directions are abbreviated using the first letter of the word. A dash between two compass directions means "to", as N-S, north to south. We often use the word "gut". This is the local term for a streambed and the ravine it cuts.

Use the odometer on your vehicle to follow the mileages for the tours. Odometers vary from one vehicle to another by as much as 0.1 mile per mile. Therefore the mileages you note on your odometer may not exactly match those given here. As you follow a tour, periodically mark your own odometer readings in this book and it will help you find the points of interest listed. The readings we list are rounded to the nearest 0.05 mile.

Some things will have changed since this book was written. Especially subject to change are the unpaved roads. Unless we have said otherwise, an unpaved road is passable by two-wheel passenger car. If there has been a major storm recently, an unpaved road may be in poorer condition than we have described. Such damaging storms are most likely to occur from late August to early October, coinciding with the hurricane season. Drive carefully and slowly on rough roads. For short distances on poor roads it is best to walk. Certainly if a road seems worse than we have described, and you are not in a suitable vehicle, turn back.

Maps to the entire island are on the last two pages of this book. Although there are several other maps in this guide, you will find it helpful to have a copy of a large detailed road map of St. Croix. An excellent map is the "Official Road Map" available free at the Visitor's Bureaus in Christiansted, Frederiksted, and the airport. It is sometimes in short supply, but another good map is "Map of Maps", also free at several locations including vehicle rental agencies, Visitor's Bureaus, and larger hotels. Topographic maps, highly detailed and showing most of the mills and ruins, are sold at V.I. Photo Supply [see p. 18, mile 0.1(---)].

There are three main kinds of transportation available to the tourist on St. Croix. There are standard taxis, which are required to carry a fare card of the fixed rates between oft-traveled points. The most traveled route, between the two main towns, Christiansted and Frederiksted (described below under "Centerline Tour"), is traveled by van taxis, which pick up and drop off passengers all along the way. The fare between any two points is $1. To hail one of the vans, simply stand on the side of the road and flag the driver with your hand. Rental cars from three major agencies are available at the airport. Other rental agencies are in Frederiksted and Christiansted, including several that rent four-wheel drive vehicles.

Some suggestions: Handy items for touring St. Croix are camera, film, drinking water, binoculars, sunscreen lotion, a detailed map, and a pencil (for odometer notes). Wasp nests, common in vegetation and ruins, are not easily noticed [picture, p. 111]. Bump a nest and you will be stung, so be vigilant when walking off the road. If you are birdwatching deep in the mangroves, you may need insect repellant. Polarized sunglasses are very useful for seeing into the water. Some reef features described are those detected by color changes in the water, which are more clearly seen with polarization. Theft from parked cars is a problem. If you are away from your vehicle, don't leave anything of value in it. Remember to drive in the left lane.

As you travel about St. Croix, please respect private property. We spoke with many property owners as we gathered information for this guide. Most of them were not opposed to allowing visitors to take a short hike over their undeveloped property (forest, pastureland, etc.) unless the area was posted for no trespassing. If in doubt, ask. Private property that was tacitly open to the public at the time we inquired, may no longer be. Owners may have changed their policy, or ownership may have changed. Be considerate. Don't leave trash and don't be destructive. Be gentle around old ruins; they crumble. Avoid damaging the ruins or yourself. Regardless of what you may hear or signs you may see, all beaches in the Virgin Islands are open to the public from the vegetation line down to the water. A beach can still be privately owned, but you have the right to be on it. You do not have the right to be inconsiderate; leave no trash, follow any posted beach rules, and if you must play your radio, do it quietly.

HISTORY

Some doin' well, some seein' hell.
--old Cruzan saying

If you want to see an area with a colorful past, choose the Caribbean. And if you want to see history played out on the land in front of you, come to St. Croix. It started with the Indians, at least four different cultures, sweeping up through the West Indies from South America. They lived and fought here leaving tantalizing evidence of their presence throughout the island. When Columbus reached St. Croix in 1493, the Caribs were apparently in the process of eliminating the Arawaks. Columbus sent a landing party for water, which also helped themselves to things in the Indians' houses while the owners were out. The Caribs, correctly sensing they were being visited by hooligans like themselves rather than civilized people, battled the sailors, but lost.

Over sixty Indian settlements have been located on St. Croix, but by 1600 the Spaniards, without starting any settlements of their own, had cleaned the Indians off the island. But the Indians enjoy a posthumous laugh, having introduced everybody to tobacco. English, French and Dutch all tried permanent settlements in the early 17th century, using slave labor from Africa since there were no Indians left. In the mid-17th century the English kicked out the Dutch and French. Then the Spaniards shipped over from Puerto Rico and booted the English. The Dutch then tried to return, thinking the Spaniards had gone home to Puerto Rico. They had not and blew the Dutch out of the water. Then the French came back and drove out the Spaniards.

Next St. Croix was purchased from the French and for three years it was owned privately by a man named de Poincy. He had a grandiose plan to dominate the commerce of a large collection of Caribbean islands. He controlled all the export-import operations of St. Croix, bought low and sold high, taxed heavily, and became rich and disliked. A revolt of the colonists nearly ensued and de Poincy quickly gave the island to the Knights of Malta in 1653, taking a fat tax write-off against profits he had from various French islands. Up to this time St. Croix produced more tobacco, ginger, and indigo than anything else. Under new management, St. Croix began major sugar production, with Dutch merchants taking more and more control. But disease, droughts, and financial feudings among French and Dutch merchants, planters, and politicians induced the King of France to step in, forcing the Knights of Malta to sell St. Croix to the Crown in 1665.

About this time France declared war on England, presumably to let the English know they weren't welcome in Caribbean commerce. Then in 1667 the English, Dutch, and French agreed to follow island ownership as it existed just before the war, regardless of who had captured what. The next year England and Denmark made some commerce agreements, and Denmark sailed into the West Indies, as if the islands needed more partitioning governments. Denmark claimed St. Thomas and St. John, since they were among the few islands nobody else wanted. In 1673 wars began again, but for variety it was France against Holland and Spain, and England against everybody. Various islands changed hands, but St. Croix remained French, carved into 90 plantations, with nobody doing very well due to wars and trade restrictions. By 1696 the French were fed up with the financial losses of St. Croix and shipped their colonists and slaves off to Hispaniola.

For the next 37 years St. Croix, still owned by the French, was left to hangers-on. In 1733 Denmark bought the island since it was their nearest neighbor to the south, and they did not want to chance an unfriendly nation taking it over. A wise move since, by this time, you couldn't trust anybody. The Danish West India and Guinea Company, a combined venture of private interests and the Danish Crown, surveyed St. Croix and delineated the plantation boundaries which are still used today, essentially unchanged. (Currently land titles are listed as numbered plots of the original Danish estates.) The Company sold plantations to anybody, intending to make money on duties and shipping, rather than production of crops. Planters came from nearby English, French and Dutch islands, but Danes were not keen on leaving Denmark. The English became dominant in number, so that, even though Denmark owned the Virgin Islands for two centuries, English is our language and we are still driving on the left.

By the mid-18th century St. Croix was a major agricultural site, complete with smuggling, rebellious slaves, and much carping about the trade restrictions imposed by the Company (shades of de Poincy). In 1755 the Danish Crown bought out the West India Company, thus ending the last private control of St. Croix. During the next 70 years the sugar economy boomed as never before. This was the period of windmills, factories, and elegant greathouses. But farming then was as it is now, some hit it and some were clobbered.

In the 1770s Europe carried on its wars at home and throughout the West Indies, while the American Revolution further complicated life. Denmark was ostensibly neutral, so the merchants and farmers of St. Croix did their best to subvert the myriad wartime trading restrictions by smuggling whatever they could to the highest bidder, evading taxes, privateers, and pirates. For profits, there is nothing quite like selling to all sides in a war. The volume of shipping throughout the Caribbean was unbelievable. Some of the plantation ruins you will see are the remainders of that wartime opulence. By 1782 war in the West Indies ended again.

After a relatively quiet 20 years, the English began the usual island grabbing. In 1801, by anchoring a large force offshore from the Danish Virgin Islands, England convinced Denmark to surrender them. This was accomplished without shooting; so much for the security provided by all the forts and batteries around the islands. A year later the islands were returned to Denmark, who knows why. Quiet until 1807, then Napoleon pressured Denmark out of her neutrality to ally with France. Naturally the English couldn't let that slide and declared war on Denmark. They promptly took the Virgin Islands in the same quiet way they had done in 1801, again avoiding, as the English request for surrender stated, "the unavailing effusion of blood". Too bad wars aren't still like that. This time the English stayed in command for 8 years. That didn't really bother most of the residents since almost all plantations were owned by English anyway. After the Napoleonic Wars England returned the Virgin Islands to Denmark. It was the least they could do, the Danes had been such good sports in the non-battles of '01 and '07.

St. Croix continued to experience the usual fluctuations of its sugar-based economy until the mid-1800s. At that time perfectly good sugar started coming from expanding plantings of sugar beets, which could be grown in temperate regions. That addictive crystal was no longer an exotic tropical luxury; Europe and the U.S. could produce it at home.

Furthermore, although owning slaves was legal, their importation to the Virgin Islands had been outlawed in 1803. As the slave labor supply line was cut off, imagine what happened to the price of slaves existing on the island. It might have been cheaper to pay wages. With the cost of labor up and crop prices declining, farmers chose the three usual routes: get out, go broke, or modernize. Modernization meant steammills with their much reduced labor requirement over windmills. The chimneys you can see are the marks of the steam era. By the last quarter of the 19th century the windmills were silent. There had been about 150 in operation, and they had whirled over St. Croix for more than a century.

In an attempt to phase out slavery, Denmark proclaimed in 1847 that thenceforth all those born to slaves were free and in 12 years all slaves would be free. We don't need to tell you what the planters thought of abolition. In 1848, 8000 slaves from the west end of St. Croix, not seeing any good reason to wait 11 more years, marched into Frederiksted, demanded immediate freedom, threatening to fire the town. The Governor-general of the Virgin Islands, Peter von Scholten, long a champion of abolition, rushed to Frederiksted. In a masterful stroke against bureaucracy, he stood at the fort and without the authorization from his King or anyone else, summarily freed all Danish West Indian slaves, about 20,000 on St. Croix. Freedom was, no doubt, unbearably good news to the slaves, but one very important item was omitted. Payday. Payday for a lifetime of work, for generations of work. Picture the chaos that followed when suddenly 90% of the population was without money, property, or a job; farmers who could give them jobs were irate over the loss of their slaves, and trying to farm a crop that was doomed.

The Labor Act of 1849 meant to establish economic order. This Act set wages at 5 to 20 cents a day (depending on skill), plus housing, plus an area for personal garden, minus 5 cents a day for food from the company store. Almost everybody was designated as unskilled, earning 5 cents a day. Use some quick mathematics and you'll get the idea. The economy of St. Croix continued to flop, prices dropped, population fell. Neither the plow nor steam could save sugar. The laborers tolerated this farce called emancipation for 30 years. In 1878 they rioted for a week burning a great deal of St. Croix west of Christiansted: cane fields, sugar factories, greathouses, schools and much of Frederiksted. Fifty-one estates were destroyed. Some of the ruins you see today have been abandoned ever since that fateful week. Thus was born the Labor Act of 1879 which allowed for negotiable wages. Two years later the government began giving land to laborers. The majority of farmers continued to pursue the phantom profits of sugar and Denmark monotonously tried to bail them out with money and regulations.

Sale of the islands to the United States had been periodically discussed since 1865, sometimes nearly culminated. Finally in 1917 the U.S. bought the Danish Virgin Islands for 25 million dollars. The main purpose was to prevent Germany from acquiring them, giving her a base for New World naval operations for World War I. One month after the transfer the U.S. and Germany were officially at war. The Virgin Islands were administered by the U.S. Navy until 1931, when they were transferred to the Department of Interior as a Territory. The U.S. Government continued to prop up the islands as the Danes had done, while the Depression made a bad farming situation worse. The government started a homestead program, buying land for 10 to 30 dollars an acre and selling 10 acre parcels on long term mortgage to people without land. Everyone still looked to farming as a salvation.

In 1936 the Virgin Islands received its constitution, and for the first time everyone of legal age could vote in local elections (although not for governor who was appointed by the U.S. President). A voter was no longer required to be male and moneyed. In 1970 Virgin Islanders could elect their own governor. We still cannot vote in national elections, but it's not a case of taxation without representation. All the income tax money collected (at Federal rates) is kept by the Virgin Islands. The last stalk of commercial cane was harvested in 1966. In that decade tourism became a notable part of the economy, and by the 1980s has become the basis of the economy.

Travel with this book in hand for a tour through 500 years of history and natural history. Stand amidst the ruins of mill or greathouse and comtemplate the miseries and the ecstasies that transpired on that very spot. You can be an Indian, Columbus, a planter, a pirate, and a slave--the past is palpable, for this is St. Croix.

Fig 3. Bodkin Mill, a scenic outlook, a link with the past.

This section is designed especially for the cruise ship passengers who arrive in Frederiksted in the morning and must depart that afternoon. The tours outlined here could also be used by other day-trip visitors starting from a point other than Frederiksted, such as the airport or seaplane port. If you do not wish to rent a vehicle, ask a taxi driver what his charge would be to drive you around on a tour described below, or what he would charge to drive you around for a given number of hours. For cruise ship passengers the closest vehicle rental is Midwest Rent-a-car across the street and to your right half a block after you exit from the pier. Their hours are M-Sa: 8-12am and 1-5pm.

CENTERLINE TOUR: Frederiksted to Christiansted. Highlights: fort, ruins, museums, botanical garden, rum distillery tour. Be selective, you can't see all of these well in a day. The tour starts at the Frederiksted post office p. 41. Read up the page in the column (W-E).

TOUR OF FREDERIKSTED: Highlights: fort and other historical sites, beaches, snorkeling, SCUBA. If you are starting from the cruise ship, you won't need a vehicle. The tour starts on p. 60, but if you are just exiting from the pier, start on p. 64, mile 0.6(1.0) at the pier and then choose your own direction.

WEST SHORE TOUR: (10 miles round trip) Highlights: beaches, snorkeling, ruins, scenic headland with lighthouse.
1. Start from Fort Frederik and drive north out of town on Route 631, which joins Route 63, West End Road, 0.25 miles north of the fort.
2. Find that junction of Routes 63 and 631 [p. 72, mile 2.2 "West End Road"] and follow the guide going S-N.
3. At Ham Bluff, simply follow the guide in reverse (N-S) back to Frederiksted, remembering to continue straight S where Route 63 turns inland at mile (4.75).

FOREST DRIVE (9.6 miles round trip, includes 2.95 miles graded dirt): Highlights: beaches, horseback trail rides, tropical forest, wood crafts. This circle tour includes parts of West End Road, Creque Dam Road, and Mahogany Road.
1. Start from Fort Frederik and drive north out of town on Route 631, which joins Route 63, West End Road, 0.25 miles north of the fort.
2. Find that junction of Routes 63 and 631 [on p. 72, mile 2.2, "West End Road"] and follow the guide going S-N.
3. Turn right at mile 3.7, onto Route 58. [Go to p. 88, "Side Trip Along Creque Dam Road", and follow the guide W-E.]
4. Turn right at mile 2.95, onto Route 765 and follow the paved road to the junction with Route 76. [Route 765 is discussed as a branch off "Side Trip Along Mahogany Road", p. 83, mile (6.4)]. Turn right onto Route 76 and follow the guide (E-W) reading up the column from mile (6.4).
6. At the end of Route 76 [p. 81, mile (9.65)] turn left onto Route 63 and return to Frederiksted.

SCENIC DRIVE (18.5 miles round trip, includes 7.7 miles graded dirt): Highlights: tropical forest, wood crafts, mills and ruins, sweeping vistas of the north, south, and west shores from the highest road on St. Croix, views

of northern Virgin Islands. This circle tour includes all of Mahogany Road and parts of River Road, Scenic Drive West, Creque Dam Road and West End Road.
1. Start from Fort Frederik and drive north out of town on Route 631, which joins Route 63, West End Road, 0.25 miles north of the fort.
2. Find that junction of Routes 63 and 631 [p. 72, mile 2.2, "West End Road"] and follow the guide going S-N.
3. Turn right at mile 2.4 (0.2 mile later), onto Route 76, Mahogany Road. [Go to p. 81, mile 0.0, and start W-E of "Side Trip Along Mahogany Road".]
4. At mile 5.45, Route 76 ends and you will then be on Route 69, River Road.
5. At mile 6.25 go straight N past the turnoff of Route 72, and you will remain on Route 69. [Find this junction on p. 135, mile 3.15(1.25), under (S-N), in "Side Trip on River Road".] Follow the guide from this point reading up the (S-N) column.
6. At mile (3.6) turn left onto the dirt road, Scenic Drive West, Route 78. [Find this junction or Routes 69 and 78 in "North Shore Tour" on p. 114, mile 11.7 under E-W.] Follow the guide from this point reading down the E-W column.
7. At mile 16.15 turn left onto Creque Dam Road (Route 58). [Find it on p. 92, mile (0.0), under (E-W) in "Side Trip on Creque Dam Road".] Follow the guide from this point reading up the (E-W) column.
8. At mile (3.2) is the junction of Routes 58 and 63. Turn left onto Route 63, West End Road. [Go to p. 72, mile (3.25) under (N-S) of "West End Road".] Follow the guide from this point reading up the (N-S) column.
9. At mile (4.75) on Route 63 is a junction with Route 631. Continue straight and you will be on 631 returning to Frederiksted.

BUCK ISLAND: See p. 179. Take a taxi between Frederiksted and Christiansted.

EAST END AND SOUTH SHORE TOUR (58 miles round trip includes 4.7 miles graded dirt): Highlights: beaches, excellent snorkeling, vistas, mills and ruins, lagoons, saltponds, East Point. There are more than a day's worth of things to see and do on this tour. Spend time on what interests you the most and save the rest for a week-long stay on St. Croix. Keep track of the time to allow yourself a relaxed drive back to Frederiksted. If you are east of Christiansted, allow two hours for driving time to return to Frederiksted; if west of Christiansted, allow 1 hour. With normal traffic it takes 75 minutes to drive from East End to Frederiksted. This tour includes Centerline Tour, Northeast Shore Drive, South Side Road, and South Shore Road.
1. Start "Centerline Tour" from the Frederiksted Post Office [p. 41, mile (0.0)]., and read up the column (W-E). The tour ends near the Christiansted fort.
2. Start East End Road, Route 82, from near the Christiansted fort, [p. 137, mile 0.0, "East End Tour"] and read down the column W-E.
3. After you reach East End, backtrack 2.3 miles to the junction of Routes 66 and 60 [p. 159, mile 0.0], "Route 60, East End Tour"]. Turn left onto Route 60, Southshore Road, and follow the guide from mile 0.0 under E-W.

4. At mile 6.3 turn left onto Route 624. [Directions 4 and 5 continue with the "East End Tour".]
5. At mile 6.9 turn left onto Route 62, Southside Road.
6. At mile 14.3 the tour ends. Join Centerline at that point by continuing straight or N [p. 25, mile 3.5, under E-W, "Centerline Tour"], and return to Frederiksted.

Fig 4. If you look up into trees even occasionally, you will certainly notice one of these mounds of dark brown material. The majority of these termite nests belong to the arboreal **Long-nosed Termite**. These insects are responsible for an enormous amount of clean-up of vegetational detritus and litter of paper and other wood-based products. They construct their nests and tracks (the dark line leading down the tree trunk) of a mixture of feces and saliva. The tracks sometimes range as much as 100 yards from the nest. The tracks keep the termites in the moist dark as they pass between their feeding areas and the nest. If you remove a section of track you will first see fat-bodied workers disoriented in the opening. In a short time, sleek soldiers, which can inflict a noticeable bite, arrive to protect the workers as they repair the damaged track. The Tree Lizards seem to know the difference, preferring to eat the workers and leave the soldiers.

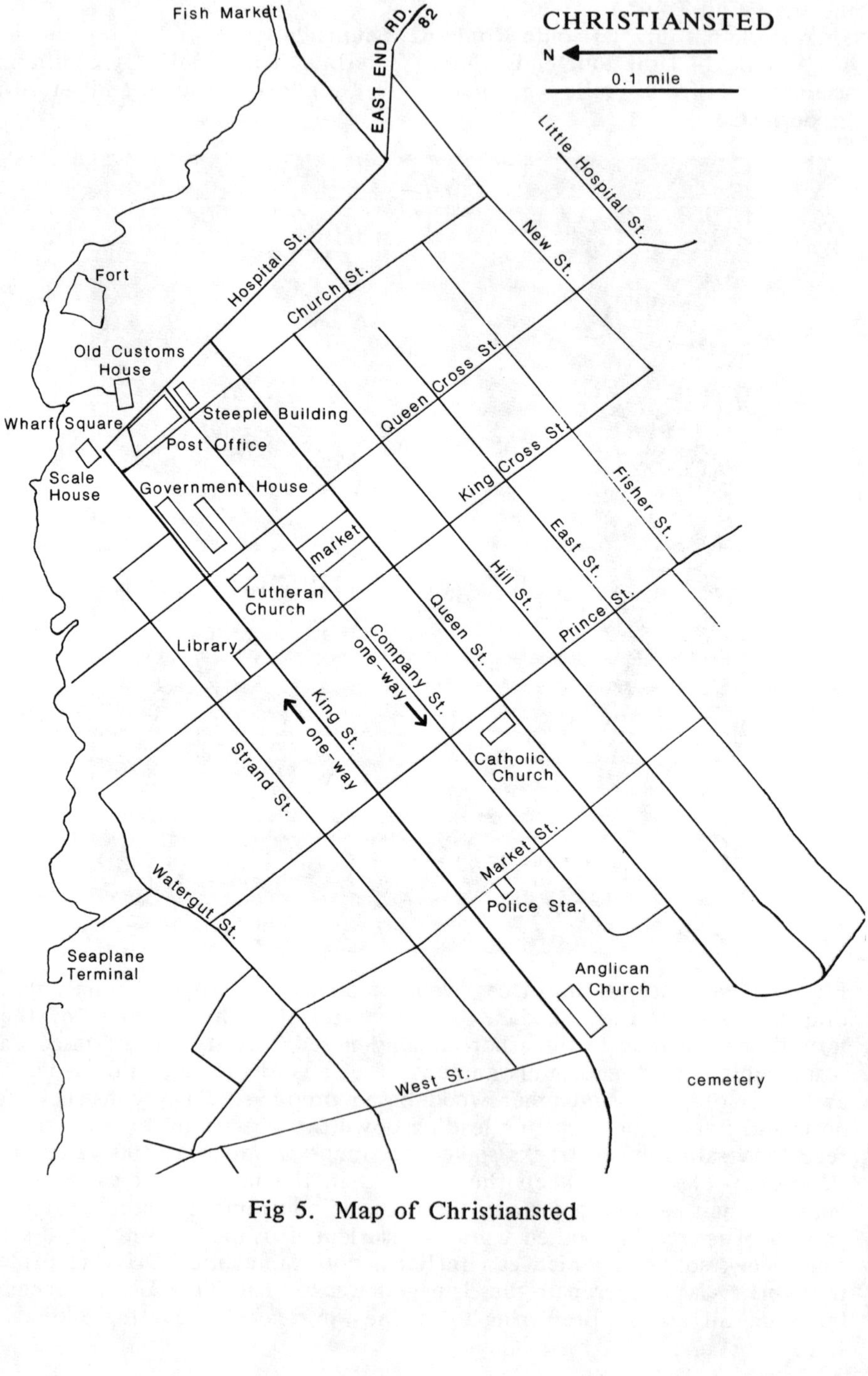

Fig 5. Map of Christiansted

CENTERLINE TOUR from CHRISTIANSTED to FREDERIKSTED

INCLUDING ROUTE 70 and PARTS OF ROUTES 75 and 753

This tour is along the major artery from Christiansted to Frederiksted, the two towns of St. Croix. It is the busiest road on the island and goes by the three major shopping centers. It is the route taken by public transportation (van taxis). Principal attractions are ruins and other historical sites and good hilltop viewpoints. Roads off Centerline provide access to birdwatching areas. Initially, it includes sections of Routes 75 and 753.

Christiansted Tour:
E-W (W-E)
0.0 (15.4) The start of the tour is the northeast end of Company Street at intersection with Hospital Street (Route 82) S of the parking lot of **Fort Christiansvaern.** Park somewhere nearby and walk around the historic site.

-The Fort is open M-F 8-5, Sa-Su 9-5. Construction of most of the fort took place in 1738-1749. Present restoration duplicates its appearance in the mid-1800s, after an extensive renovation period in 1834-1837. Brochures and signs allow one to take a leisurely self-guided tour through the structure, from dungeon to officers' quarters. It is now the headquarters for the Christiansted National Historic Site and the Buck Island National Monument. The gift shop sells guide books. There are **public restrooms** inside the Fort.

-**The Steeple Building** on the SE corner was a Lutheran Church constructed in 1750-1753, the steeple in 1793-1796. It is part of the Christiansted National Historic Site and is used as a **museum**. One of the most interesting displays is a **collection of potsherds and other pre-Columbian artifacts collected on St. Croix.**

-The National Historic Site includes the **Old Danish Customs House and Post Office** on the SW corner of the Fort parking lot, and the buildings that are now the **U.S. Post Office** and **U.S. Customs** directly S across the street from the Old Customs House. This **complex of buildings was built by the Danish West India & Guinea Company** (which created the plantation structure of St. Croix--375 estates of 150 and 300 acres size). There are **public restrooms** in the old West India complex.

-One can arrange for **tours to Buck Island and other water sports** including SCUBA diving and fishing around the Wharf Square parking area and along the waterfront to the W of the parking lot. Most of the commercial vessels dock in this area.

-A **ferry** leaves every 10 minutes from here for **Protestant Cay**, the islet just offshore in the center of the harbor. There is a $3 charge for persons not staying at Hotel on the Cay. Protestant Cay was named during the French administration in the late 1600s when non-Catholics could not be buried on the main island. This island is one of two places in the world (the other is Green Cay National

E-W (W-E)

Wildlife Refuge off northeastern St. Croix) where one can find the endemic **St. Croix Ground Lizard**. They are the large, blue-gray lizards seen on the ground (not in the trees). This lizard was once widespread along the strand areas of St. Croix, but was extirpated in the early 1970s. The introduced mongoose probably ate the lizard to extinction on St. Croix, taking 100 years to do it. (There are no mongooses on Protestant and Green Cays.) A project is underway to extirpate the mongoose from Buck Island National Monument and release this rare lizard to that island, where it probably once occurred.

-Directly W across the Wharf Square parking lot from the Old Customs House is a relatively modern building, the **Old Scale House**, built in 1855. The old scale is still operable. Stand on it and notice the great mechanical advantage; one child can balance the weight of several adults. This building houses a **government Visitor's Bureau**, which gives out pamphlets describing walking tours of Christiansted and Frederiksted.

-Company Street is one way, southwest, out of town. The first three blocks of Company Street are lined with **many old buildings in good repair**, dating from the 1700s and early 1800s. (When doing this tour backwards, W-E, you will end with King Street at the harbor parking lot [Wharf Square] just west of the Fort.)

Fig. 6. View of Protestant Cay from Fort Christiansvaern.

Company Street (Companiesgade), one-way southwest:
<0.1>
0.1 (---) Begin the tour by driving along Company Street past the Steeple Building. V.I. Photo Supply, at the intersection with Dronnigen's Tvergade, sells **U.S.G.S. topographic maps** of the islands. Telephone: 773-1055.

Christiansted Tour

E-W (W-E)
<0.05>
0.15 (---) **Open air market** on left between Dronnigens Tvergade and Kongens Tvergade. Private vendors of food and many other products are selling here every day. It is a good place to get an idea of local produce and talk with West Indians regarding use of herbs and other things they sell.

<0.15>
0.3 (---) Holy Cross Roman Catholic Church on left at intersection with Prince Street. This Church was built in 1755 and extensively remodeled in the 1850s. A right turn here goes to the seaplane terminal, 0.2 mile.

<0.05>
0.35 (---) End of Company Street (Companiesgade) part of the tour. Turn right or NW onto Market Square (or Street). A police station is on the left or W. [Note that the next 4 entries (down the page) apply to those doing the tour (W-E). To continue the tour going E-W, skip to mile 0.4(15.0).]

- -

End of King Street (Kongensgade), one-way northeast, at Wharf Square.

<0.1>
--- (15.3) **Government House** on right, just across Dronnigens Tvergade from the Lutheran Church. There is an impressive stairway leading up to the ballroom of the Danish governors. This building has housed Danish governors and government offices (continuing today) from the time construction was begun in 1747. The peaceful **interior courtyard** was landscaped during the Depression by the Civilian Conservation Corps. It is shaded by some very large **West Indian Mahogany Trees** festooned with the bromeliad **Old Man's Beard**. There are several **Lignumvitae Trees**, three near the central pool. One about 15 feet high at the NE corner of the fish pond has peeling gray and black bark that reveals the underlying green layers. **Public restrooms** are located on the E side of the first floor. You must usually supply your own toilet paper. [If going (W-E), continue 0.1 mile to Wharf Square and park. Note that the next 3 entries (up the page) apply to Company Street. To see the Company Street sites park somewhere in the Wharf Square vicinity and skip to the descriptions at mile 0.0(15.4).]

<0.05>
--- (15.25) **Lutheran Church** on right at the intersection with Dronnigens Tvergade was originally built in the 1740s as a Dutch Reform Church. It was purchased and remodeled in the 1830s by the Lutherans, and the congregation from the Steeple Building moved here. Many of the interior furnishings, including the floor tiles, were moved here from the older Steeple Building.

<0.1>

Christiansted Tour

E-W (W-E)

--- (15.15) **St. Croix Public Library**, Christiansted Branch on the left at the intersection with Kongens Tvergade. The library has **public restrooms**.

<0.05>

--- (15.1) Limpricht Park on the right or E. In summer the yellow flowers of several **Golden Shower Trees** near the fence bordering King Street are hard to miss.

Fig. 7. The courtyard at Government House in Christiansted has Lignumvitae and Mahogany Trees.

<0.1>

0.4 (15.0) King Street (Kongensgade) and Market Square (Street). If going towards Frederiksted, turn left or SW onto King Street. (For those traveling E-W towards Frederiksted, to see the King Street stops you will have to turn right here and follow the mileage stops upward from mile ---(15.1) to 0.0(15.4) at Wharf Sqaure. Then retrace your steps to this point.)

[Coming from Frederiksted (W-E), continue straight or NE, one-way on King Street to the Wharf Square parking area at the Harbor. There are **numerous fine old buildings dating from the 1700s and 1800s** on both sides of King Street. The next 4 mileage stops (up the page) are on King Street. The stops above them with --- in the (W-E) mileage column indicate stops on Company Street.]

E-W (W-E)

<0.05>

0.45 (14.95) On the left or SE is the impressive St. John's Anglican **rectory and church, made of limestone blocks**, 1849. The gothic-revival style of the sanctuary is greatly enhanced by curious pinnacles which adorn its roof and have ornamental projecting bricks, called "crockets".

<0.1>

0.55 (14.85) Road to left or SE borders the **Christiansted cemetery**, a large area with many very old gravestones and monuments. At 0.2 mile on the SE corner of the intersection is the **former island hospital at Peter's Farm** (1855-1950s), now privately owned and not a hospital.

<0.1>

0.65 (14.75) The **Fredensthal Moravian Church** on the left or S dates from 1852 and was built to hold 1000 worshipers!

<0.1>

0.75 (14.65) Bassin Triangle traffic light. Proceed straight or W to continue the tour.

-A left or S turn at Bassin is a 0.8 mile shortcut (Contentment Road) out of the urban area of Christiansted and rejoins the tour at 1.85(13.55) below.

-A right or N turn goes to the **seaplane terminal** (airboat, Goose)

Seaplane Terminal (S-N):

0.0 Junction at Bassin Triangle traffic light

0.2 A left or W turn goes 0.1 mile to the **ruins of the greathouse of Estate Richmond**. Other outbuildings are nearby. Drive across the intersection at the stop sign onto a dirt driveway to see the ruins.

0.45 Driveway to seaward for **seaplane**

<0.15>

0.9 (14.5) Golden Rock **Post Office** on right or N

<0.25>

1.15 (14.25) Paved road to right or N. The **old Danish prison** (used until the early 1970s and now abandoned) is on the NW corner of an intersection at 0.15 mile. It was built in the Danish Colonial Style in the 1840s for 100 men, and is said to be modeled after the layout of the stables of Estate Richmond. This structure was considered a great reform over the dungeons at Fort Christiansvaern. This part of town also had the insane asylum and a leper colony.

<0.05>

1.2 (14.2) Paved road to left or S leads to **greathouse and mill of Estate Orange Grove**. The road is lined with West Indian Mahogany Trees.

<0.2>

1.4 (14.0) Junction with start of Route 75 at traffic light. Golden Rock **Shopping Center**, which includes a **supermarket**, is

E-W (W-E)

on the SW corner. Go straight or S onto Route 753 to continue the tour. Turn right or W to begin Route 75. [The log for North Road, Route 75, is on p. 93.]

<0.45>

1.85 (13.55) Junction of Route 753 and Contentment Road at stop sign. Continue straight or S. A turn to the left or E would take you back to the Bassin Triangle traffic light [p. 21, mile 0.75(14.65)].

<0.2>

2.05 (13.35) Road to left or E to Questa Verde condominiums and an **excellent view of Christiansted area**:

Hermon Hill Road and Questa Verde (W-E):

0.0 Junction with tour at mile 2.05(13.35). Watch for speed bumps ahead.

0.3 Road to left or N to private residence built among **mill and ruins of Estate Hermon Hill**

0.55 Questa Verde Condominiums on the left. Take the left fork here.

1.2 Three-way fork. Go straight onto middle road, which is dirt.

1.3 Turn right onto a grassy dirt road as the main dirt road veers left. The grassy road requires four-wheel drive, so park here or at pavement end if you have two-wheel drive. A large tree, heavily laden with Spanish Moss, is visible at the junction on the left side of the grassy road. The grassy road climbs about 200 feet in 0.25 mile, but the **panoramic views** are worth the 10-minute walk.

Fig. 8. View of Christiansted and Protestant Cay from Jacobsberg Ridge.

The crest of this hill is **Jacobsberg Ridge** (which includes Recovery Peak above and SE of you). Just as you reach

View from Jacobsberg Ridge (continued)

the crest you get the best views of the south shore. From left to right:

-Halfpenny Bay, just right of hill with a white-domed house.

-Manchenil Bay, with the **mill and ruins of Estate Diamond Keturah** on an inland rise at the Bay's midpoint.

-Vagthus Point (Danish) or Watchho (English), the elevated peninsula forming the western boundary of Manchenil Bay

-Canegarden Bay is inbetween Watchho and the Hess Channel, with the greathouse of Estate Canegarden on a knoll at its midpoint.

-Hess Channel, often with large tankers at dock.

Now walk out to the northern end of this hilltop, slightly downhill about 150 yards. From here you can see from Watchho and further W to continue from the Hess Channel.

-Ruth Island, artifically created by dumping dredge spoil on a coral reef when the Alucroix Channel was made

-Krause Lagoon remnant, inland from Ruth Island.

-Hess Oil chimneys (black and silver striped) and domed tanks of Martin Marietta Aluminum refinery (now closed).

-And on a clear day: Long Point; the Bight of Long Point; Westend Saltpond; and Sandy Point, the western tip of the island.

-Inland from Long Point, the southern part of the mountains include St. George Hill with numerous antennas and dishes on top, and a high 923-foot peak just north, Quassia Hill.

-The chimney at Bethlehem Old Works sugar factory is in the valley, below and just N of St. George Hill.

-The mill of Estate Bodkin is a tiny thimble on the ridgeline just south of Eagle Ridge.

-The highest part of St. Croix, Eagle Ridge, with its two tallest peaks, Blue Mountain (with antennas, 1096 feet) and Mount Eagle (1165 feet)

-The peninsula that extends out to sea in the west is Estate Judith's Fancy.

-Proceeding east from Judith's Fancy, one can see **Christiansted Harbor**, formed by the barrier coral reef called **Long Reef**.

-The power plant and large green water tanks are **facilities of the Water and Power Authority** (called by its acronym WAPA) which provides all electrical power for St. Croix. It also makes 2.5 million gallons of potable water daily from seawater. It is built on the ruins of Estate Richmond, which had the first communal sugar factory on the island, begun and terminated in 1877. A deep green, dredged channel runs from the mouth of Christiansted Harbor to the WAPA dock. Fuel oil is barged in from Hess Oil by this route.

View from Jacobsberg Ridge (continued)

-Towards you from the large green tank of WAPA is a roofless ruin in a grove of trees, the former greathouse of Estate Richmond.

-The **seaplane terminal** is at the eastern end of a large clearing.

-**Christiansted Harbor mooring area** with many boats.

-Proceeding N out to sea: yellow **Fort Christiansvaern, Protestant Cay** and the eastern end of Long Reef. The northern Virgin Islands are visible on a clear day.

-Submerged **Round Reef**, the bogy for entry into Christiansted Harbor, is between the end of Long Reef and Fort Louise Augusta.

-Gallows Bay is on the eastern end of Christiansted Harbor with marina, piers, and numerous warehouses.

-The **mill of Estate Mount Welcome** is on a hill E of Gallows Bay.

-East of Gallows Bay is the point that has **radio station WSTX** built on the ruins of Fort Louise Augusta.

-In line with the radio station is submerged **Scotch Bank** offshore, a dark brownish green surrounded and riddled by the lighter turquoise blue of sand bottom.

-**Altona Lagoon** is the large mangrove which lies between the radio tower and the Buccaneer Hotel, a large pink structure on a hill to the east.

-In the distance are the islets Green Cay and Buck Island. Pull Point is in line with Buck Island viewed from this hill. It has several large stone buildings, and it is joined to the main part of St. Croix by an isthmus.

-The sandy island beyond Pull Point is called Candle Reef, here partly obscured by the hill inland from Pull Point.

-Just east of Candle Reef one can see the westernmost end of Tague Bay Reef

-Coakley Bay Condominiums are the tall buildings on shore inland from the end of Tague Bay Reef.

-The inland hills belong to the Seven Peaks Range. The highest is Peak Seven, 860 feet.

-The top of a navigational tower on Recovery Hill is just above you to the SE.

E-W (W-E)

<0.05>

Centerline Road, Route 70 (also Queen Mary Highway):

2.1 (13.3) Junction with beginning of Route 70. To continue the tour turn right or W. Route 83 continues straight or S for 2 miles and eventually joins Route 62. [A description of Route 83 is on p. 173, mile 11.7(2.6).]

<0.15>

2.25 (13.15) Right or N to Beeston Hill residential area. A private residence is built into the **mill and ruins of Estate Beeston Hill** on left or W at 0.1 mile. The remainder of the road is residential.

Including Route 70 and Parts of Routes 75 and 753

Centerline Road, Route 70

E-W (W-E)

<0.05>

2.3 (13.1) Junction with Route 708. Veer left or S to continue Centerline Road. A right or N turn rejoins Centerline at mile 3.9(11.5). At a distance of 0.75 mile one can turn off on a side trip to the **U.S.D.A. Experimental Forest of Mahogany and Teak Trees**. [See "Side Trip to the Experimental Forest," p. 43.]

<0.15>

2.45 (12.95) Gravel road to left or S to private **Estate Grange greathouse**. The road has many old **Lignumvitae Trees** along its sides. At 0.1 mile on the right a lawn and broad stairway lead to a monument to Rachael Faucett Levine, the mother of Alexander Hamilton. Her gravestone has not been found; the monument was placed about 80 years ago by a biographer. Hamilton, the American statesman, was on St. Croix 1765-1773 (age 10 to 18), and he lived for a while at this house. His mother, Rachael Levine, was married and buried at this greathouse, which was owned by her sister. The owners allow one to walk up the lawn and steps to visit the monument, <u>not</u> the greathouse. South of the Levine marker is a monument to Danish soldiers who died during a yellow fever epidemic in 1886. The greathouse was an isolation ward. For a while later, Grange was a botanical garden experiment station, and a nice collection of plants is still present. Behind the soldiers' memorial is a Danish bell dating from 1711.

<0.1>

2.55 (12.85) Paved road to right or N into Grange Hill Nursery. Hours are M-Sa 8am-5pm. Telephone is 773-7092.

<0.45>

3.0 (12.4) Paved road to left or S to **mill and ruins of Estate Anna's Hope**. Just before an elaborate peaked bell tower on the right at 0.1 mile is a tree with very large leaves, a **Teak Tree**. On the rest of the grounds are numerous old buildings and the remains of horticultural plantings including a collection of palm trees. Make continuous right turns to complete a 0.3 mile loop through the main part of the ruins and return to Route 70 again about 30 yards W of where you started.

<0.5>

3.5 (11.9) Junction. **Chimney and ruins of Estate Peter's Rest** on the SE corner. To continue Centerline turn right or N. To start the South Shore Drive (Route 62) turn left or S [p. 174, mile 14.3(0.0)]. Straight ahead or W is Route 682, which runs 0.4 mile to Route 68 [p. 53, mile 2.5(0.35)]. You may visit some of the **ruins of Estate Peter's Rest** via a road off Route 682.

CENTERLINE TOUR from CHRISTIANSTED to FREDERIKSTED

Including Route 70 and Parts of Routes 75 and 753

Centerline Road, Route 70

E-W (W-E)

Route 682 to ruins and mill of Estate Peter's Rest (E-W):

0.0 Junction with Routes 70 and 62.

0.3 Turn right or N onto paved road, just before intersection with Route 68.

0.5 Turn right or E uphill. At 0.55 mile as the road veers S, on the left is a stockpile of utility poles lying among **ruins, bell tower, and mill of Estate Peter's Rest**. One can **view the surrounding area** when the bush is cut.

<0.4>

3.9 (11.5) Junction with Route 708. To continue on Centerline veer left or W. Cars rounding the corner on Route 70 have the right-of-way, but use caution because some drivers do not yield as they should.

-The building on the SE corner here was one of eight **schoolhouses** constructed in the traditional style used on St. Croix around 1840 when the Danish made public education for all (slave as well as free) mandatory. Governor von Scholten decreed that instruction would be in English in deference to the language of the majority of the islanders. Nearly all of these buildings with their distinctive style and sturdy construction are still in use, but only one of them is now a school.

-One can take Route 708 to visit the U.S.D.A. Experimental Forest. [See "Side Trip to the Experimental Forest," p. 43, under "Remainder of Route 708", mile 1.35(0.0).]

<0.45>

4.35 (11.05) Junction with Route 81. Route 81 to the left or S goes 0.65 mile to join Route 68 [p. 53, mile 2.45(0.4)]. At 0.25 mile, a road to right or W goes into private residence built around **ruins and two mills of Estate Castle Coakley**.

-One can take Route 81 to right or N to visit the U.S.D.A. Experimental Forest. [See "Side Trip to the Experimental Forest," p. 44, mile 3.15(0.0), under description of return to Centerline.]

-One can also take Route 81 for 1.95 miles to Route 74, eventually joining Route 75. [See p. 96, mile 1.95 of Route 81 and mile 1.3 of Route 74.]

<0.3>

4.65 (10.75) At the Red Cross Center a paved road to left or S becomes dirt after 50 yards. On the SE corner here is a small masonry structure with an arched roof. This is one of the "**baby houses**" used for a day nursery by the slave mothers. The unique configuration of these buildings makes them easily recognizable.

-The dirt road leads through the impressive, castle-like **ruins of Estate Castle Coakley** at 0.2-0.45 mile S of Route 70. Some of the ruins are inhabited. While wending

E-W (W-E)

through the ruins, choose the right turn at all forks, and you will emerge onto a paved connecting road to Sunny Isle Intersection. Turn right or N on this road, 0.45 mile to the shopping center. Go straight or N through the traffic light to continue the tour.

<0.1>

4.75 (10.65) Sunny Isle Intersection (with 7 traffic lights). Junction with Four-lane Highway or Route 66. Sunny Isle **Shopping Center** with a **supermarket** is on the north side of this intersection. To continue Centerline Road make a right or N turn. To begin Route 66 go straight ahead or W [p. 46]. A left or S turn goes 0.65 mile to Route 68, ending at the border fence of Hess Oil Refinery [p. 53, mile 0.65(0.0)]. Some of the **ruins of Estate Castle Coakley** are visible from this road to left or E at 0.35 mile.

4.75- 5.0 (10.4- 10.65) Sunny Isle Shopping Center to right or E. The road veers left or W again inside the shopping center. Along the northern row of buildings in this shopping center, starting near the sign that says "Mini Mall" is a line of **Sausage Trees.** They are in fruit and flower in July and August. The flowers are pendulous, bizarre, and large, and the fruits of varying shapes often look like giant sausages.

5.0 (10.4) Junction with Route 79 to right or N [p. 98]. Route 79 goes 2.35 miles to Route 75, passing the St. Croix **Hospital** and **Island Center for the Performing Arts**.

<0.3>

5.3 (10.1) Turn right or N to continue Centerline.

<0.1>

5.4 (10.0) Turn left or W to continue Centerline. A dirt road to right or E goes to a private home built among the **mill, chimney, and ruins of Estate Diamond**.

<0.4>

5.8 (9.6) Centerline turns right or N and joins Route 709. Continuing straight or W goes to St. Ann's Church, built among the **ruins and mill of Estate Barren Spot**.

<0.15>

5.95 (9.45) **Mills of Estate Strawberry Hill** are visible beyond the housing development on the right or E. Turning right or E opposite Hummingbird Lane of Barren Spot Village, one can wend through the housing to the base of the water mill, which is unusual in having an elaborate exterior stairway. Some houses have ruins of the old Estate in their yards. The old greathouse was recently torn down to make way for a church.

<0.15>

6.1 (9.3) Centerline turns left or W. Route 709 leads straight or N, 1.7 miles to Route 75.

E-W (W-E)

Route 709 goes N, then W, to Route 75:

0.0 Junction with Centerline at N end of Strawberry Hill housing complex

0.7 Two dirt roads to right or E. Continue Route 709 by going straight or N. The left dirt road passes beside a private residence on the left or W built among the **ruins and chimney of Estate Mary's Fancy** at 0.25 mile.

0.9 Veer left or W. The partially paved residential road straight ahead or N has views of the Salt River Valley with the mill of Estate Windsor perched above. A view of the chimney of Estate Mary's Fancy is on a hill to right or E at 0.1 mile.

1.1 Private residence built among **mills and ruins of Estate Bonne Esperance** on the right or N.

1.7 Junction with Route 75 [p. 101, mile 5.3(0.85)]

<0.6>
6.7 (8.7) As you come down out of the cut through a hill, on the left or SW ahead is a water mill of Estate Clifton Hill. The greathouse and a cane-grinding windmill are on the hill to the SW, mostly shrouded by trees.

<0.3>
7.0 (8.4) Junction with Route 75, the main route to the north shore. Villa LaReine **Shopping Center** with a **supermarket** is on the NE corner. "Reine" is pronounced "rain". Just south of the supermarket is a row of four **India Rubber Trees.** They are characterized by 4-12 inch glossy leaves. They grow rapidly and produce quantities of reddish roots which sprout from the branches and elongate toward the ground. These root and eventually grow into supporting trunks. Native to India, it was the first species used for commercial production of rubber. It was later replaced for rubber production by a species from Brazil which gave greater amounts of latex at an earlier age. The Brazilian rubber tree cannot be commercially grown in plantations in the New World tropics because of natural pests, but does well in the Eastern Hemisphere far removed from its enemies. India Rubber Trees are popular ornamentals in the tropics and subtropics.

-La Reine **Police Station** is adjacent to the shopping center near Route 75. [Route 75 is described on p. 101, mile 6.15(0.0).]

<0.4>
7.4 (8.0) Junction with Route 663. On the NW corner is the Kingshill Lutheran Church with its decorative chain-and-pillar fence, built in 1912.

-The road to right or N labeled "To 707" goes 0.7 mile to Route 707 and then connects to Canaan Ridge Road, which goes to the north shore [p. 133, mile 4.7(0.0)]. At a short distance away it passes **ruins and mills of Estate Slob.**

E-W (W-E)

-The road to left or S joins the Four-lane Highway (Route 66) near the industrial complex. A side road leads to a **view of the industrial complex along the South Shore**.

Route 663 to S:

0.0 Junction with Centerline at mile 7.4(8.8)

0.15 Paved road to left or E goes to private residence built around **mill and ruins of Estate Clifton Hill**.

0.8 Paved road to right or W (Ramirez Road), just before traffic light at Four-lane Highway, goes into housing complex of Profit and leads to a **view of the South Shore**.

- 0.0 Junction with Route 663 at mile 0.8.
- 0.2 Turn right or N on paved road.
- 0.35 Turn left or W uphill (just before passing beside a baseball field).
- 0.45 Government buildings amid **ruins and mill of Estate Profit**. When the center is open, from the mill and courtyard there is a good **view of the southern shoreline** including Martin Marietta Alumina Plant; remnants of the island's largest mangrove, Krause Lagoon, seaward of Martin Marietta; and Ruth Island, which was made by dumping dredge spoils from the Alucroix Channel onto a coral reef.

0.85 Junction with Route 66 at a traffic light [p. 50, mile 2.15(4.05).] To the left or E is an experimental planting of **Hybrid Mahogany Trees**. Continuing straight or S through the traffic light one comes to the main gate of Martin Marietta.

<0.1>

7.5 (7.9) Dirt road to N runs among inhabited **ruins of Estate King** (included in Upper Bethlehem). It eventually joins Route 707 at the southern end of Route 73, Canaan Ridge Road, but the last part is over limestone marl which is badly eroded and requires four-wheel drive.

<0.1>

7.6 (7.8) Road to left or S with a sign on the corner "Sheltered Workshop". Drive 0.15 mile uphill. The walled compound on the right or W is Griggs Home for the Aged in the **old Kingshill Barracks**. The grounds include a tomb of a Danish soldier. Drive through the pillars into the central area and park. Just past the pillars is a good-sized **Lignumvitae Tree** on the right and and a **Gumbo-Limbo Tree** on the left. The **old Danish schoolhouse** at the northern end of the compound is used by the **Sheltered Workshop, which sells handsome basketry and other craft items** made by the handicapped of St. Croix. They will make items to order. The shop is open M-F 9am-3pm.

Including Route 70 and Parts of Routes 75 and 753

Centerline Road, Route 70

E-W (W-E)

-The large pink building just beyond or W of the road to Kingshill Barracks is the Virgin Islands Territorial Court.

Fig. 9. The Sheltered Workshop at the old Kingshill Schoolhouse sells basketry and other craft items made by the handicapped of St. Croix.

<0.15>

7.75 (7.65) Junction with Route 707 to right or N. One can use this road to visit **Fredensborg Pond** (freshwater) and **ruins of Bethlehem Old Works Sugar Factory**--see estate£ nearby. The visit to these sites is described below.

Fredensborg Pond and Bethlehem Old Works (circle, basically E-W):

0.0 Junction with Centerline at mile 7.75(7.65)

<0.65>

0.65 Junction. Turn left or W here onto dirt to go to the Bethlehem Old Works Sugar Factory.

-Continuing straight or N is a dirt road that goes 0.05 mile to **Fredensborg Pond and the old water mill of Estate Fredensborg**. This is the largest and only year-round freshwater marsh on St. Croix. It affords **good birdwatching**. If you take this track, at 0.05 mile, turn left towards the pond, drive around the pond edge, and up onto the dam. At 0.2 mile you will be in the center of the dam and due south of an old water mill.

-Route 707 turns right or E here to run 0.8 mile to Route 75 [p. 101, mile 5.55(0.6).]

Fredensborg Pond, Bethlehem Old Works, E-W (continued):

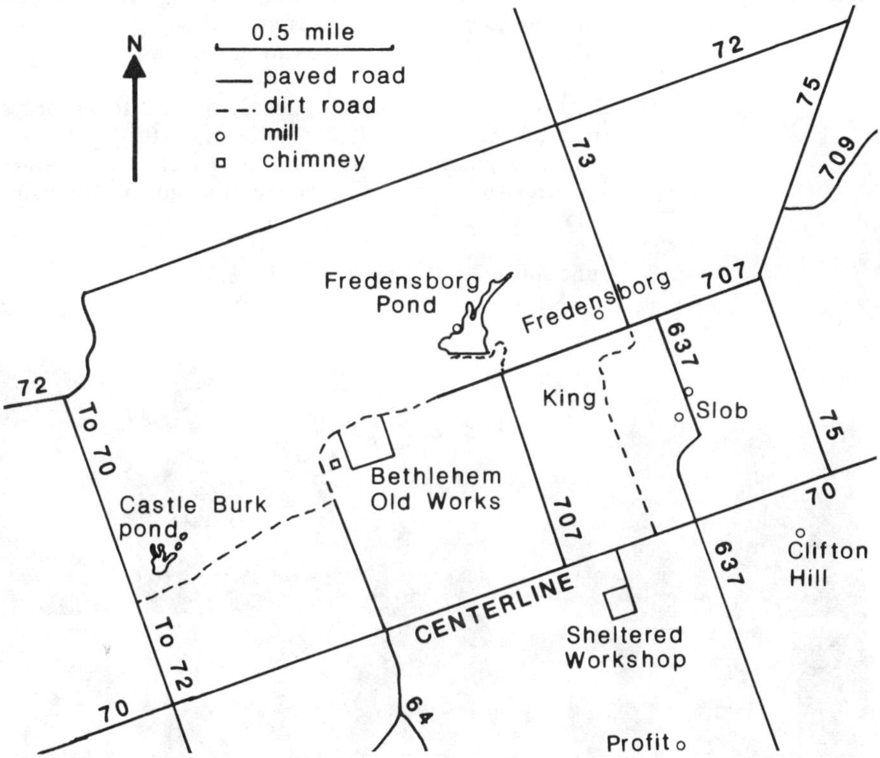

Fig. 10. Map of Fredensborg and Bethlehem Old Works area.

Remainder of Route 707 (W-E):

1.05 Junction and start of Route 73 to left or N. Overgrown **ruins and mill of Estate Fredensborg** are on the NW corner. The dirt road taking off to the SW at this junction goes over an eroded track (four-wheel drive) past inhabited **ruins of Estate King** (also called Upper Bethlehem) to end at Centerline [p. 29, mile 7.5(7.9)].

1.15 Route 663 to right or S. This road runs by Fredensborg Housing project and **ruins and mills of Estate Slob**, ending in 0.7 mile at Centerline [p. 28, mile 7.4(8.0)].

1.45 End of Route 707 at Route 75 [p. 101, mile 5.55(0.6)].

Continuing to Bethlehem Old Works (E-W):
<0.45>

1.1 On the left or S are the **Ruins of Bethlehem Old Works Sugar Factory**, the last sugar factory on St.

Fredensborg Pond, Bethlehem Old Works, E-W (continued):
Croix, which closed during the 1960s. The 207-foot chimney is visible for miles around.
-The road gradually curves to left or S.

<0.3>

1.4 Dirt road to right or W goes 0.65 mile to an access road of Route 70 [p. 87, mile 0.9(0.05)]. At 0.55 mile one passes among **ruins of Estate Castle Burk**. Turn left or S on the access road to go to Centerline, 0.05 mile [p. 33, mile 9.0(6.4)].

<0.4>

1.8 Junction with Route 70 at mile 8.35(7.05)

Fig. 11. Fredensborg Pond and old watermill of Estate Fredensborg.

<0.6>

Continuing Centerline Road:
E-W (W-E)

8.35 (7.05) Junction with east end of Route 64 (Airport Road). Turning left or S goes to the **airport** and two of the few remaining **mangroves** on the island. [See "Side Trip Along Airport Road," p. 54 for a description of this route.] Airport Road makes a large "U", rejoining Centerline at mile 11.1(4.3) [p. 36].
-The U.S. Department of Agriculture (U.S.D.A.) Experimental Station and the University of the Virgin Islands have **fields of tropical crops** planted on the SW corner of this junction. The entrance is 0.25 mile S on Route 64.

<0.5>

8.85 (6.55) Paved road left or S to **University of the Virgin Islands Extension Campus**. The **old greathouse and some**

E-W (W-E)

outbuildings of Estate Golden Grove are now occupied by U.O.V.I. Extension Service on top of the western hill. The roadway is lined with Caribee Royal Palms (South American Royal Palms). Many other **labeled plant specimens** are on the grounds, especially on the western hill.

-Across Centerline from the entry are labeled **plots of tropical forage crops**.

<0.15>

9.0 (6.4) Junction with paved road. A right or N turn on "To 72" goes 0.95 mile to Route 72 [p. 87, mile 0.95(0.0)]. A left or S turn goes to Golden Grove **Police Station** at 0.6 mile and ends at 0.9 mile on the Four-lane Highway [p. 51, mile 4.2(2.0)].

<0.2>

9.2 (6.2) A right or N turn goes into the **Virgin Islands Government Agricultural Station**, open M-F 8am-5pm. There are **experimental fields of tropical crops**. The grounds have a hybrid mango orchard, a banana grove, a citrus orchard, and many kinds of **vegetable crops**. Tropical fruit trees are for sale. The annual **Agricultural and Food Fair** is held here at the station in mid-February and is worth visiting if you are interested in West Indian food, crafts, and agriculture. There are exhibits from several Caribbean countries.

<0.15>

9.35 (6.05) Dirt road to right or N just beyond or W of a gas station. This is the old cobble road leading to ruins of **Estate Lower Love** (also called Estate Die Liefde). At 0.15 mile veer left onto a field road. The **chimney and ruins** are at 0.25 mile amid heavy growth. Lower Love was transformed into a modern steammill at the end of the 19th century, serving as a central factory for processing sugar cane from many estates. The tall chimney created a powerful draft of air through the combustion chambers. It is 112 feet high, the walls are 6 feet thick at the base, 18 inches thick at the top; the base is 24 feet square on a foundation that is 11 feet deep.

<0.05>

9.4 (6.0) The square **chimney** to right or N is part of the **ruins of the factory at Estate Lower Love**. In the background one can see the **highest peaks of the island comprising Eagle Ridge**. The nearer and lower peak on the E with all the antennae and radio dishes is Blue Mountain, elevation 1096 feet. The higher peak to the NW is the highest point on the island, Mount Eagle, elevation 1165 feet.

<0.4>

9.8 (5.6) Junction with paved Route 69. A right or N turn onto Route 69 goes to the North Shore (4.4 miles). At 1.1 miles

E-W (W-E)

distance it connects with Route 72-76 [p. 135, mile 4.4(0.0)].

-The trees on the NW and SE corners of this intersection are **Indian Almonds**. These trees have large, shiny leaves, and a distinctive growth pattern with the smaller branches projecting horizontally from the trunk. The "almonds" are edible. This tree is a native of the East Indies and Oceania, but is widely naturalized in the tropics.

-A left or S turn connects to the Four-lane Highway, Route 66.

Route 69 connection to Route 66 (N-S):

0.0 Junction with Centerline at mile 9.8(5.6)

0.4 Paved road to left or E to private government buildings (the residence of the Lieutenant Governor) among **mill and ruins of Estate Adventure**. The mill is above on the hilltop.

0.6 Paved road to right or W. Less than 10 yards further on the left or E is a grassy track to **mill and ruins of Estate Negro Bay** (includes Estate Fox). The walk to the start of the ruins is about 0.2 mile and to the mill another 0.2 mile. The grassy road to E eventually veers N. At this veer, walk south into bush. Wend through brush past ruins of workers homes (rebuilt in this century) until you come to another dirt road. Cross this dirt road through a border of Snake Plant (characterized by variegated, strap-like leaves about two feet long). The mill is 50 yards south of the Snake Plants. The houses of the village are arranged in two rows, each with about 6 houses. Each house is separated from the others by about 50 yards. In what must have been the village center is a **huge Pencil Tree**. This village layout is typical of the plantations established in the early 1700s with individual houses for the slave families. Later construction economized on costs and time by using long houses in an apartment arrangement. Estates Anguilla [p. 55, mile 0.3] and Hogensborg [p. 37, mile 12.2(3.2)] have examples of the long houses.

-An easier walk to the mill uses a dirt road just south of the first one. At the first fork go right. At the second fork go left. At the third fork go left past a house ruins with blue-green walls on your right. The next ruins are on the left and a border of Snake Plant is on the right. The mill is about 50 yards south of the Snake Plants.

0.8 Turn right or W following sign "To 66".

1.2 Government housing project built around **ruins and mill of Estate Paradise**

E-W (W-E)

Route 697 connection to Route 66, N-S (continued):
1.7 Traffic light junction with Route 66 at mile 5.7(0.5) [p. 51]

<0.6>
10.4 (5.0) Junction with Route 705. A right or N turn goes 1.1 miles to Route 72-76.

Route 705 (S-N):
0.0 Junction with Centerline at mile 10.4(5.0)
<0.2>
0.2 **Ruins and mill of Estate Plessen** to right or E, sometimes hidden by bush.
<0.65>
0.85 On the left or W are the **chimney and ruins of greathouse of Estate Grove Place.** The greathouse is between the Grove Place Fire Station and a paved road to government housing project. The road veers to E or right.
<0.05>
0.9 Junction. Turn left or N to continue to Mahogany Road.
<0.05>
0.95 On the left or W is a **huge Baobab Tree** which has split into three trunks. St. Croix's first labor union was formed in 1916 and convened many meetings under the shade of this tree.
<0.1>
1.05 On a concrete structure on the left or W side of the road is a **mural "Tribute to Robert Nesta Marley"** by Paul Youngblood and apprentices. Bob Marley was a very popular Rastafarian singer of Reggae, a musical style from Jamaica. The mural has numerous symbols which relate to Rastafarian beliefs.
<0.05>
1.1 Junction with Mahogany Road [p. 85, mile 5.3(4.35)].

<0.05>
10.45 (4.95) Dirt road to left or S leading past St. Joseph's Chapel on the SW corner. At 0.15 mile a heavily overgrown dirt road to E (hiking only) goes to ruins of **greathouse of Estate Mount Pleasant.** This greathouse was considered elaborate, but the ruins are now almost hidden.

<0.3>
10.75 (4.65) Road to right or N goes to **St. George Village Botanical Garden** (phone 772-3874). The grounds are open dawn to dusk all days; the gift shop operates M-Sa 9am-3pm. The nursery operates T,Th,Sa 9am-noon. This 16-acre garden was acquired, planted, built and restored, and continues to run through donations and the efforts of volunteers. The site is the former **sugar factory and village of Estate St.**

E-W (W-E)

George, and the ruins have been used as a base for the present buildings or are in various stages of restoration. The collection has about 300 species of **tropical plants, horticultural and native**, and 225 specimens have been labeled throughout the grounds. A map is available at the main building. On Thursday mornings there are demonstrations in the old blacksmith shop, as iron decorative items are created for the gift shop.

Fig. 12. The blacksmith shop at St. George Botanical Gardens has demonstrations on Thursday mornings as ironwork is created for the giftshop.

<0.35>

11.1 (4.3) Junction with the western end of Route 64 (Airport Road). The distinctively-styled building on the SE corner here is an **old school house** built in the 1840s, now used by the Boy Scouts. It is open most weekdays. Go in and enjoy the airy roominess that the trussed ceiling gives.

-Airport Road makes a broad "U" to the E, passing the **International Airport** and rejoining Centerline at mile 8.35(7.05) [p. 32]. The **Cruzan Rum Distillery** gives **tours** at its plant, 0.65 miles S of this junction. The **mangroves** of Manning Bay Lagoon and Krause Lagoon are also reached by this road. [See p. 57, mile 5.75(0.0), for a description of this road.]

-The road to right or N is residential. At 0.4 mile on the right or E is a **picturesque old water mill of Estate St.**

E-W (W-E)

George. One can drive 1.35 mile total, ending at a locked gate outside ruins of **Estate Bog of Allan** (also called Allandale). The **mill of Estate Clairmont** can be seen through the gate on a hill to the north.

<1.1>

12.2 (3.2) Pond Lane. A left or S turn goes into a residential area. A right or N turn onto a dirt road goes 0.15 mile to some **very fine ruins of Estate Hogensborg**, a large plantation of the 1800s with a huge three-story greathouse, overgrown and picturesque. The entry road passes some long houses (slaves' quarters) on the left. The greathouse on the right was enlarged on the north side when Hogensborg served as a sugar factory in the early 1900s. The foundation remains of the manager's house are near the SE corner of the compound. After viewing Whim Greathouse (restored) at mile 13.4(2.0) below, it is fun to mentally reconstruct what Hogensborg must have been like. On the grounds there are old Banyan Trees, a very large cycad, several large palms, and numerous outbuildings with some machinery of the factory remaining. Hogensborg had the first steammill on the island. Periodically a colony of **Fruit Bats** occupies the portion of the greathouse that still has a roof.

Fig. 13. Hogensborg has extensive ruins including an elaborate greathouse.

<0.15>

12.35 (3.05) Dirt road to right or N goes up St. George Hill and provides an **aerial view of the southwestern coast of the island**. A left or S turn through the Estate Carlton pillars goes to Long Point **mangrove**.

Right or N on dirt road to top of St. George Hill:

0.0 Junction with Centerline at mile 12.35(3.05)

0.8 Three-way fork. Take the right or uphill road.

0.9 View of the vertical beds in the Judith Fancy Formation on left, an impressive **rock exposure** made by a quarry. This formation is the main component of the highest peaks on St. Croix. Gray rock on the northern edge is the Cane Valley Formation.

1.3 Fork. Go left to the top of St. George Hill.

1.7 Top of St. George Hill. This is part of the Atlantic Fleet Weapons Training Facility [details on p. 73, mile 3.9(3.05)]. Park and walk to the hilltop above the fenced compound. Visible from west to east (right to left):
-The Frederiksted Pier and red Fort Frederik
-Westend Saltpond and Sandy Point beyond
-Brown streak in ocean about 0.5 mile offshore is effluent from Cruzan Rum Distillery (visible when distillation is in progress).
-Just west of Long Point, which is the prominent point just slightly SE of St. George Hill, is the dark green area of Long Point Mangrove. It is bordered by Manchineel Trees on the inland side.
-Dense housing is Williams Delight
-Alexander Hamilton International Airport, with runway and line of buildings to south of runway
-Shoreline of airport area is dark green mangrove of Manning Bay Lagoon
-Ruth Island is a dredge spoil island based on part of a coral reef. Inland from the island a row of dark green outlines the remnants of a large mangrove area, Krause Lagoon, much of which is now covered by the industrial complex.
-Chimneys to east are Hess Oil Refinery stacks.
-Raised headland visible beyond the industrial complex is Vagthus Point (Danish name) or Watchho (most common name used by the residents). This point and some cliffs to the west of it are nesting sites for White-tailed Tropicbirds in the winter.
-Tallest chimney (round) in center inland is among ruins of sugar factory at Bethlehem Old Works
To the N, from east to west (right to left):
-Towers and microwave units on Blue Mountain (elevation 1096 feet)
-Flat-topped Mount Eagle (elevation 1165 feet, the highest point on St. Croix)
-A radio tower on the scenic road

Including Route 70 and Parts of Routes 75 and 753

Centerline Road, Route 70

View from St. George Hill (continued)
- -Bodkin mill on a ridgeline
- -Northern Virgin Islands (on a very clear day)
- -Mills of Estates Mount Victory and Nicholas barely visible at the skyline
- -Mill of Estate Punch just below the skyline
- -Puerto Rico to the NW (on a very clear day)

You can get a clear view of the SW by walking 150 yards along a level path leading east from near the entrance to the station.

Route 70 (continued)
E-W (W-E)

12.35 (3.05) (continued) Left or S through the Carlton pillars onto paved road to **Long Point mangrove birdwatching area**

Long Point Mangrove "To 701" (N-S):

0.0 Junction with Centerline at mile 12.35(3.05)

0.05 On right or W a partially paved road goes 0.3 mile to deserted hotel built around **mill and ruins of Estate Carlton**

0.8 Junction with Route 701 to right or W

0.95 Junction with Four-lane Highway

1.2 Veer left or E

1.4 Veer right or S

1.8 Turn right onto dirt road

1.9 Park as dirt road veers N. Walk about 0.2 mile W towards shore to reach Long Point Mangrove. White, Black, and Red Mangrove Trees are all found here along with saltponds.

<0.75>

13.1 (2.3) Route 703 is a paved road to S and dirt to N. At 0.15 mile to the left or S is a **water mill of Estate Whim** on the right or W.

<0.3>

13.4 (2.0) **Whim Plantation Museum** on left or S is the headquarters of the St. Croix Landmarks Society. This **restored greathouse, mills, and plantation works** create a feeling for the grand past life of the planters of the period from the 1700s to the middle 1800s. The museum covers the history of the Virgin Islands from Columbus' landing to the purchase by the United States in 1917. Guided tours of the greathouse are given all days 10am-5pm. Admission is $3 for adults, $1 for children (worth it!) The gift shop sells guide books and historical literature as well as gifts. Among the interesting historical items sold are copies of the Oxholm map of 1793 that shows the plantation boundaries and populations at that time. Telephone: 772-0598.

<0.2>

13.6 (1.8) Junction with Route 63. A right or N turn begins Route 63, which bypasses the main section of Frederiksted and then

E-W (W-E)

goes north along the western coast. [See "West End Road", p. 71.]

Fig. 14. Whim Plantation Museum has a reconstructed animal mill or whim; cane-grinding windmill; and steam plant.

<0.4>

14.0 (1.4) Route 701 to right and left. One can visit the **mill of Estate Camporico** (also called Estate West End), which is unusual in being built primarily out of bricks rather than stone and coral blocks. Go left or S on Route 701. Turn left or E onto Route 632 at 0.4 mile. The mill is on the right or S at 0.1 mile further.

<0.15>

14.15 (1.25) Dirt driveway to abandoned **greathouse of Estate Hannah's Rest**, extensively modernized

<0.2>

14.35 (1.05) **Police Station** on left or SE corner of paved cross street

<0.05>

14.4 (1.0) Centerline or Route 70 veers right or N at this junction. Cars traveling around the corner on Route 70 have the right-of-way, but use caution when approaching cars coming in from the south or west. They do not always yield. To continue Centerline into Frederiksted, veer right. A left or S turn onto Route 661 goes to **Westend Saltpond, a fine birdwatching area**, and **Sandy Point, a lovely two-mile stretch of sandy beach** where huge **Leatherback Turtles nest**. [See "Side Trip to Sandy Point", p. 57.]

-Proceeding straight across to W one can go 0.85 mile to the quiet east side of Westend Saltpond, a good vantage point for **early morning birdwatching**, with the rising sun behind you.

E-W (W-E)

Road to eastern side of Westend Saltpond (E-W):

0.0 Junction with Centerline at 14.4(1.0)

0.05 **Ruins of factory of Estate Stony Ground** to left or S, heavily overgrown.

0.4 Pavement ends

0.6 Road deteriorates to four-wheel drive only

0.85 End of road at Westend Saltpond

<0.4>

14.8 (0.6) Left or W into Smithfield residential area. At 0.25 mile, the Smithfield Inn (restaurant) is built around the **ruins of the mill of Estate Smithfield**. This mill is unusual because it is made of small coral blocks.

<0.6>

15.4 (0.0) End of Centerline Road and Route 70 at **Frederiksted**. A **U.S. Post Office** is on the SE corner. [To continue through Frederiksted to Route 63, the road along the west shore, see "Tour of Frederiksted, and Connecting Link To Route 63," p. 61, mile 0.0(1.6).]

Fig. 15. The Gray Kingbird, a Caribbean native, sings "pitirree" from open perches.

Fig. 16. The Crapo or Giant Toad was introduced to St. Croix in the 1930's to control cane insects, an experiment that did not work.

Fig. 17. Teak grove in the Experimental Forest.

SIDE TRIP TO EXPERIMENTAL FOREST via ROUTE 708

The U.S. Department of Agriculture planted the Experimental Forest of Mahogany and Teak Trees to examine the possibility of growing timber crops on St. Croix. St. Croix provided the testing grounds for U.S.D.A. experiments in a dry tropical environment. Several kinds of trees in the Mahogany Family are represented as well as their hybrids. The forest is usually deserted, peaceful, and shady.

E-W (W-E)
0.0 (3.5) Junction with Route 70 at mile 2.3(13.1) [p. 25]
<0.75>
0.75 (2.75) Paved Road to right or N. Turn right to go to the Experimental Forest. Veering left or S, you may return to Centerline on Route 708, 0.6 mile.

<0.1>
Remainder of Route 708:
E-W (W-E)
0.85 (0.5) Paved road to left or S as Route 708 veers right or W
<0.5>
1.35 (0.0) Junction with Route 70 at mile 3.9(11.5) [p. 26]

<0.15>
0.9 (2.6) Private house amid **mill and ruins of Estate Constitution Hill** are to the left or S.
<0.15>
1.05 (2.45) **View of Martin Marietta and Hess Oil** on south shore
<0.85>
1.9 (1.6) Behind a gate on right or N is the southern boundary of **U.S. Department of Agriculture Experimental Forest.** A **grove of Pacific Mahogany Trees** is behind the fence. This tree is a native of Central America from southern Mexico to Costa Rica. It never did really well on St. Croix because it was especially susceptible to pests like mealybugs and the diseases they carry.
<0.6>
2.5 (1.0) Junction with end of Route 811 at the locked entrance to Estate Sion Hill. The ruins of this estate are not visible from these roads but can be seen perched on a hill in the distance from Route 81, mile 0.9 [p. 96]. To visit the interior of the Experimental Forest, turn right or N here. The following describes the return to Centerline, by making a left or S turn here.

Return to Centerline, via Routes 811 and 81.
E-W (W-E)
2.5 (0.65) End of Route 811 at locked and gated entrance to Estate Sion Hill
<0.3>
2.8 (0.35) **Water mill of Estate Sion Farm** is on the left or E

Side Trip to Experimental Forest via Route 708

E-W (W-E)

Return to Centerline, via Routes 811 and 81 (continued):
E-W (W-E)

<0.1>

2.9 (0.25) Route 81 turns off to right or W. Continue straight or S to return to Centerline. Route 81 runs 1.8 miles to end at Route 74. On the NW corner of this intersection is a school built among the **mill and ruins of Estate Sion Farm**. [See p. 96, mile 1.8, for a description of this intersection and Route 81.]

-Turning left or E here one can drive 0.45 mile to the road end. The groves of trees near the National Park Service houses are **Windblown seed experimental plots**. The parents were Honduran Mahogany, valued for its fast growth and strong wood, and West Indian Mahogany, valued for its termite resistance and deep red color. The numbered trees in the plots are hybrid trees that grew from wind-blown, cross-pollinated seed of the two different species of Mahogany Trees.

<0.25>

3.15 (0.0) Junction with Centerline Road [p. 26, mile 4.35(11.05).]

Make a right or N turn outside gate to Estate Sion Hill onto 811 to continue to U.S.D.A. Experimental Forest. "No Trespassing" signs refer to the land off the road. There is a **grove of mixed Honduran and West Indian Mahogany Trees** behind the fence of Estate Sion Hill on the left or W. Honduran Mahoganies have large leaves; the native West Indian have small leaves.

<0.1>

2.6 (0.9) Junction with dirt road to Experimental Forest. Turn right or E here onto dirt. The paved road continues 0.2 miles further to private residences in Estate Thomas, passing by **groves of Teak and Mahogany Trees** and an old cemetery inside the fence of Estate Sion Hill.

<0.4>

3.0 (0.5) Cross a cattle guard into the Experimental Forest. Just beyond or E of the fork ahead is a **Teak Tree grove** planted 1955-1958. Just left or N of the fork is a large tree with a diameter of about 3 feet and peeling red bark. It is a **Gumbo-limbo or Turpentine Tree**, also called Tourist Tree due to its bark. Take the right or S fork soon after crossing the cattle guard. The dirt roads that run N and S off the main road go to the sides and back of various groves. Some require four-wheel drive.

<0.3>

3.3 (0.2) U.S. Forest Service Field Office (closed) in the middle of a Teak Forest. An unmaintained pit toilet is in the rear, 100 feet N of the office.

E-W (W-E)
<0.05>
3.35 (0.15) A three-way fork lies just beyond or E of the Field Office. To continue the tour, take the middle fork to E. The left or N fork ends in 0.25 mile at locked gate, passing by groves of Mahogany Trees.

N or left fork:
0.0 Fork.
0.15 Pacific Mahogany Grove, 1961, to left or N. The Pacific Mahogany has very narrow leaves compared to the other two mahogany trees planted in the experimental forest.
0.2 Cattle guard and end of Forest

Continuing on middle fork:
3.35 (0.15) Along the right or S side of the road here are several lines of small **Common Lignumvitae Trees**. They are no longer common, due to the desirability of their very dense wood (so dense it sinks in water). Although this tree is native to St. Croix, we know of no certain wild specimens. In back of the Lignumvitae is a fine grove of Mahogany Trees.
<0.05>
3.4 (0.1) Fork. Take the left or upper road.
<0.05>
3.45 (0.05) **Mixed groves of Honduran and West Indian Mahogany Trees** planted for a wind-blown seed experiment, 1928 and 1965. This experiment was an attempt to produce hybrids that would incorporate the different strengths of two mahogany species. The Honduran Mahogany has stronger wood and grows rapidly, while the West Indian Mahogany has a deeper color and is more termite resistant.
<0.05>
3.5 (0.0) Cattle guard at end of Experimental Forest. One can continue on this road and exit onto North Road (Route 75) [1.7 miles away, p. 93, mile 0.35(19.15)]. <u>However</u>, the initial part of the road is over badly eroded limestone marl and requires four-wheel drive. If you do not exit here to North Road, you must return the way you came in to the gate of Estate Sion Hill [p. 43, mile 2.5(1.0)]. The description of the return to Centerline resumes at that mileage.
-A dirt road (four-wheel drive) goes right or S into the groves of the Forest just before this cattle guard. The following are details for this road through the groves.

Road through the groves (N-S):
0.0 Junction with dirt road near the cattleguard. This road initally parallels the fenceline of the Experimental Forest.
0.2 Veer right or W

Road through the groves, N-S (continued)

0.25 Fork.

Left or SE fork:

0.0 Fork.

0.25 Sign on left indicates **groves of Chittagong and Cigar-box Cedars** planted 1967, 1968. These species are in the mahogany family. Turn left after passing through gate in fence line.

0.3 Separate, labeled groves of **Chittagong and Cigar-box Cedar**. The Chittagong Tree is a native of India, used in cabinetry there. The Cigar-box Cedar is a native tree with 1-2 foot long leaves that have many pairs of lance-shaped leaflets. The sap smells like garlic.

0.35 Locked gate. Return the way you came.

Right or W fork:

0.0 Junction

0.1 Road is lined on both sides with **Sweet-lime**, which bears small, edible, red citrus fruits in summer and fall.

0.15 Junction. Make a left, W, turn.

0.20 Junction. Make a left, NW, turn.

0.25 Junction with main Forest road between the field office and entrance.

SIDE TRIP ALONG FOUR-LANE HIGHWAY

(Also Route 66, Melvin Evans Highway)

Melvin Evans was the first elected governor of the U.S. Virgin Islands. All previous governors were appointed by the President of the U.S. This road provides a fast transit (55 mph) between the center of the island and areas to the west, avoiding the heavy traffic of Centerline Road. It provides access at several points to the industrial complex on the southern shore. It is the main access to birdwatching in the remnants of Krause Lagoon and a fast way to reach the airport. A dashed entry, "---", indicates that the stop is unavailable to traffic in that direction.

E-W (W-E)

0.0 (6.2) Sunny Isle Intersection and junction with Centerline, Route 70 [p. 27, mile 4.75(10.65)]. Sunny Isle Shopping Center is on the NE corner. The road to the left or S goes 0.65 mile to the fence outside of Hess Oil Refinery on Route 68 [p. 53, mile 1.9(0.95)].

<0.7>

0.7 (---) Just before the start of a guard rail on the left or S, one can catch a glimpse of one of two mills of Estate Clifton Hill on the hill ahead to right side of the highway. It is near the highway, below the crest, almost overgrown. The mill

E-W (W-E)

of Estate Barren Spot, near a red-roofed church is almost directly on the right or N.

<0.3>

1.0 (---) **Lignumvitae** and **Turpentine Trees** in front of a concrete tank. The wood of Lignumvitae is the densest in the world--it sinks in water. The Lignumvitae trunks are brown with patches of green. In late spring the tree is covered with small, purple-blue flowers. Later, bright yellow fruits festoon the trees, and in September the protruding bright red seed of each fruit completes a colorful display. The Turpentine or Gumbo Limbo Trees have peeling, red papery bark, giving them another name, Tourist Tree. The sap and crushed leaves smell of turpentine, but commercial production is from a pine, not this tree.

-The dirt road running south from here joins Route 68 [p. 52, mile 0.65(2.2)]. It goes 0.35 mile through a forest of unusually tall **Acacia** and **Tantan Trees**. It is a quiet, shady area. Sometimes people make charcoal in this forest; the wood smolders under mounds of earth called "coal keels". If you smell smoke, a keel is cooking.

<0.4>

1.4 (4.8) Junction with Route 68, a paved road to left or S. This road travels along the outer fence of the Hess Oil Refinery and provides good views of the refinery. It is also an access road for a **very fine birdwatching area** at Canegarden Bay, mile 2.05(0.8). [See "Side Trip Along Hess Oil Fence, Route 68", p. 52.]

-On the right or N here is a dirt track which ends near one of the two **cane-grinding mills of Estate Clifton Hill**. One can park on the first corner of Route 68 to the left or S. Walk the dirt track uphill until it dead-ends, nearly due W and uphill of the mill. Follow paths through brush taking right turns to the mill. One can also reach the mill from the open field below and just east of the mill.

<0.25>

1.65 (4.55) Traffic light. A left or S turn goes to the **container port** and **good birdwatching** among the **mangroves** along the Alucroix channel. The initial part of the road cuts through a fine **geological exposure of a fossil reef**.

Alucroix Channel Road (N-S):

0.0 Traffic light

0.4 Begin fossil reef cut, exposure to right. To explore this cliff face you will have to park at pullouts either before or after the exposure. There is no road shoulder at the cut. This reef is of Pleistocene age, only several million years old--geologically very young. The coral and snail fossils are very numerous, and some are modern species, still living

Alucroix Channel Road, N-S (continued):

on the reefs around St. Croix. Some of the corals are no longer found in the Caribbean, although close relatives occur on the Pacific side of Central America. The former reef crest was near the highest point of the road cut. This separates the seaward and landward parts of the reef. Coral species and structure change dramatically in this area.

0.85 Dirt road to right or W leads to mangroves, salt flats, ocean jetty, and birdwatching areas.

Channel frontage road (N-S):

0.0 Junction with Alucroix Road

0.35 Young **Red Mangrove Trees** grow in the channel. They are distinguished by their 2-3 inch long, leathery leaves and their most distinctive feature, the arched stilt roots (prop roots) which allow them to breath in spite of being rooted underwater. The pendulous green seedlings remain on the tree until they are several inches long. The Red Mangrove is the typical plant of the seaward edge of the mangroves, usually in the water.

0.6 Dirt road to right (less than 0.05 mile to end)

0.6-1.0 **Mixed stands of all three mangrove plants** on the right. The **Black Mangrove** has salty-tasting, gray-green leaves (lick them!) and pneumatophores (fleshy projections rising a few inches out of the soil from its roots) that allow it to breathe in low-oxygen sediments. It has button-like fruits with no grooves. The Black Mangrove is the typical plant of the salt flat areas. The **White Mangrove Tree** has dull, yellow-green leaves with a peripheral area of pores on the underside. They have a slight or no salty taste. The button-like fruits have grooves. It is the typical mangrove tree on salty hummocks which stand above the salt flats. It does not have exposed roots.

1.1 Large green molasses tank (the supply for Cruzan Rum Distillery) on the left. To the right, less than 0.05 mile is a large clearing and parking area. The area left or E of the molasses tank is a **good birdwatching area of salt flats and black mangroves.**

1.2 End of the jetty. The widened end allows turning around if you are careful. Back up, if in doubt. To the right or W is **Ruth Island**, created by dumping dredge spoils onto a coral reef that was cut to make the Alucroix Channel. To the left or E is the St. Croix

Alucroix Channel Road (continued):
Channel frontage road, N-S (continued):
Container Port. Inland and in the distance over the molasses tank is Blue Mountain with its antennas.
Alucroix Channel Road, N-S (concluded)
1.9 St. Croix Container Port

Fig. 18. Red and Black Mangroves. The rootlets sticking directly upward are pneumatophores off the Black Mangrove roots. The dark thin pendulous "fruits" hanging from the branches of the Red Mangrove are actually its seedlings.

<0.25>
Route 66 (continued)
E-W (W-E)
1.9 (4.3) To the right or N is a reforestation project with **hybrid Mahogany Trees**. To the left or S inside the fenced grounds of Martin Marietta Alumina Corporation are cleared **ruins of Spanish Town greathouse**. It has an

E-W (W-E)

unusual rounded section and several decorative circular windows.

<0.25>

2.15 (4.05) Traffic light. The road to right or N is Route 663 [p. 29, mile 0.85], which goes 0.85 mile to Centerline. There is a **view of the industrial complex** from the **mill of Estate Profit**, about 0.5 mile away, described on that page.
-The road to left or S is the **main entrance for the old alumina refinery of Martin Marietta**, now defunct. The grounds are open to the public M-F 5:30-7:30am, 4-7pm; Sa-Su all day-7pm. To visit the **Spanish Town ruins**, which are the cleared ruins on the rise to the NE, turn left into the parking area just past the entrance guard house. Park here and walk to the ruins. The entrance of the parking has a grove of **White-cedar Trees**. Two examples are the trees you pass between just as you enter the compound. They have papery pink, tubular flowers and usually several trunks of fissured, gray bark.

Fig. 19. The greathouse of Estate Spanish Town has a rounded section with some decorative circular windows.

<0.9>

3.05 (3.15) Road cut on left or S exposes the **Kingshill Marl**, a geologically recent deposit (Miocene, about 20 million years ago) of limestone interbedded with marl. This deposit is widespread over the middle portion of St. Croix, producing most of the outcropping rocks between the mountains of the eastern and western ends of the island. The marl erodes easily, and the limestone forms cliffs. The limestone has been a principal source of building stone for the many old estate buildings found all over St. Croix. Although the deposit was apparently formed in deep water, in places it includes shallow coral and reef fossils which

E-W (W-E)

were apparently deposited by turbidity currents, large landslide-like flows of sediment that originated in shallow water and finally fanned out depositing their loads in deep water. It is from these fossils, many in very good condition, that the age was determined.

<0.15>
3.2 (3.0) Traffic light. Eastern junction with Airport Road, Route 64 [p. 55, mile 1.25(4.5)]. A left or S turn here takes you to Alexander Hamilton **International Airport** and access to the two largest **mangroves** on the island.

<0.5>
3.7 (2.5) Golden Grove Territorial Prison to right or N

<0.3>
4.0 (---) Road cut to S exposes the Kingshill Marl [details on p. 50, mile 3.05(3.15)]. From the top of the rise in this cut looking ahead and across the road, you can see the mill of Estate Negro Bay near the top of a low hill ahead.

<0.2>
4.2 (2.0) Junction at a traffic light. The paved road to left or S leads to a government-owned industrial park, 0.2 mile to end. From most of this road there are good views of the Negro Bay mill across the highway and to the NW. It is most easily seen when going N.
-The road to right or N goes 0.9 miles to Route 70, passing the Golden Grove Police Station.

Road to Centerline (S-N):
0.0 Junction with Route 66 at mile 4.2(2.0).
0.35 Intersection. Turn left or N to go to Route 70. The Golden Grove **Police Station** is the large building near the NE corner. A **water mill of Estate Golden Grove** is on the SE corner. A ruin is on the NW.
0.9 Junction with Route 70 [p. 33, mile 9.0(6.4)].

<1.5>
5.7 (0.5) Traffic light. Road to right or N goes to government housing project built around **mill and ruins of Estate Paradise**, and then to Centerline, 1.7 miles distant. [See "Centerline Tour", p. 33, mile 9.8(5.6).]

<0.5>
6.2 (0.0) End of Four-lane Highway at Airport Road [p. 57, mile 4.7(1.05)].

This was the terminus in mid-1986. The following is a tentative log for the new section of the highway when it opens.

0.0 (3.5) Old terminus at Airport Road, mile 6.2(0.0)

<0.2>
0.2 (3.3) **Ruins and two mills of Estate Enfield Green** are on the right or N of the highway.

E-W (W-E)

<1.95>

2.15 (1.35) Intersection with Route 703. The **mill of Estate Hope** is on the left or S.

<0.45>

2.6 (0.9) Intersection with Route 63. A **grove of mixed Honduran and West Indian Mahogany Trees** is on the SW corner.

<0.9>

3.5 (0.0) Junction with Route 661. Turn right or N, 0.6 mile to Route 70 [p. 40, mile 14.4(1.0)].

SIDE TRIP ALONG HESS OIL FENCE, ROUTE 68

(also Hope Road)

This side trip takes you along the fence of Hess Oil Refinery. It provides access to a good birdwatching area, and it bypasses the busy shopping center of Sunny Isle. It starts at mile 1.4(4.8) of the Four-lane Highway, Route 66, and ends at mile 14.15(0.15) of the East End Tour, Route 62.

W-E (E-W)

0.0 (2.85) Junction with Four-lane Highway [p. 47, mile 1.4(4.8)].

<0.05>

0.05 (2.8) Road veers left or E. A pullout is towards the fence. On the hill behind you to N is a cane-grinding **windmill of Estate Clifton Hill**. You can walk to this mill from this point by crossing the highway to a dirt track. [See p. 47, mile 1.4(4.8).]

<0.15>

0.2 (2.65) Veer right or S.

<0.2>

0.4 (2.45) Guardhouse on right or W. There are seven **Lignumvitae Trees** along the left or E side of the road near here. Several more are further along this road. The road veers left or E about 30 yards past the guardhouse.

<0.05>

0.45 (2.4) Turnout at corner for **view of Hess Oil Refinery**. This is one of the largest refineries in the western world. Frequently one can see huge tankers offshore. The refinery takes in crude oil and exports more refined oil and chemicals.

<0.2>

0.65 (2.2) Dirt road to left or N goes 0.35 mile to Route 66 [p. 47, mile 1.0(---)] through a forest of Acacia and Tantan trees.

<0.25>

0.9 (1.95) Veer right or S.

<0.3>

1.2 (1.65) Main Gate of Hess Oil Refinery as road veers to left or E. The Administration Building is just inside.

<0.4>

1.6 (1.25) Good view of the array of stacks and pipes for cracking process at Hess Oil Refinery

Side Trip Along Hess Oil Fence, Route 68

W-E (E-W)

<0.3>

1.9 (0.95) Road north 0.65 mile to junction of Route 66 and 70 at SW corner of Sunny Isle Shopping Center

Connecting Road to Sunny Isle Shopping Center from Route 68:

S-N	(N-S)	
0.0	(0.65)	Junction with Route 68 at mile 1.9(0.95)
0.25	(0.4)	Paved road into **ruins of Estate Castle Coakley**.
0.4	(0.25)	Dirt road to left or W on south side of a church. At 0.2 mile an auxillary dirt road goes left or S through many **White-cedar Trees** to the crest of a hill (elevation 143 feet) for a **view of the Hess Oil Refinery**.
0.65	(0.0)	Junction of this connecting road at Sunny Isle Intersection with Centerline (Route 70) [p. 27, mile 4.75(10.65)], and Route 66 [p. 46, mile 0.0(6.2)].

<0.1>

2.0 (0.85) On right or S, junction with road into Figtree Hill Estates at a guardhouse. To see a good specimen of a **Sausage Tree**, go 0.25 mile S to the SE corner of Limetree Drive and Scallop Lane.

<0.05>

2.05 (0.8) Dirt road to right or S to French Pond and Canegarden Bay, an **excellent birdwatching area with mangroves, salt flats, ponds, and beach**. This gate is open from 6am-6pm. [Another entry point is described with more details about the site on p. 173, East End Tour, mile 13.55(0.75)].

Dirt road to Canegarden Bay (N-S):
This road is rough when dry, muddy when wet.

0.0	Junction with Route 68 at mile 2.0(0.85)
0.95	View over saltponds to SE
1.0	Junction with dirt road to left or E. This road runs 0.35 mile on the north side of the pond mangroves. It ends near the junction with the trail from the church described at mile 13.55(0.75) of South Shore Road, Route 60-62, [p. 173].
1.25	End at seashore of Canegarden Bay. In winter, looking down the cobble-sand shore to the E, one can see **White-tailed Tropicbirds** flying near their nests in the cliffs near Vagthus Point.

<0.4>

2.45 (0.4) Junction with Route 81. You can go left or N 0.55 mile to Route 70 [p. 26, mile 4.35(11.05)].

<0.05>

2.5 (0.35) Junction with Route 682. Veering left or NE onto 682, it is 0.45 mile to Route 70 [p. 25, mile 3.5(11.9)].

W-E (E-W)

<0.35>

2.85 (0.0) Junction with Route 62 [p. 174, mile 14.15(0.15)] and end of Route 68.

SIDE TRIP ALONG AIRPORT ROAD, ROUTE 64

Airport road goes by the U.S.D.A. experimental station with numerous labeled tropical crops. It also goes near two mangroves, Krause Lagoon and Manning Bay Lagoon. It goes to the international airport. Near its western end it goes near the rum factory which gives tours of its plant.

E-W (W-E)

0.0 (5.75) East junction with Route 70, Centerline Road [p. 32, mile 8.35(7.05)]

<0.25>

0.25 (5.5) On the left or E side of the road, about 50 yards before the entrance to the agricultural fields is a large **Painkiller Tree**, a native of Southeast Asia through the East Indies and Australia, naturalized on St. Croix. Painkiller refers to the use of the large, undulating leaves in a poultice. The distinctive fruit (which resembles a skinned pineapple) is edible, but you would have to be desperate to eat it.

-Paved road to right or W into **U.S. Department of Agriculture** (U.S.D.A.) and **University of the Virgin Islands Experimental Station** (M-F, 8am-4:30pm). Turn right just inside the gate to view **experimental fields of tropical crops**. One banana grove is bordered on the E by a windbreak of Horseradish Tree and on the S by Tantan. There are plantain bananas, squat hybrid papayas, mangos, avocados, sapodilla, pineapple, breadfruit trees, grapes (!), passion fruit, and fish culture with algae for *Tilapia*, an African fish. **Other crops** include Lablab, Sorghum, Jojoba, Manioc or Cassava, Pigeon Peas, and Sugar Cane.

-The offices on the left of the entrance road at 0.05 mile are those of C.O.V.I. The road goes 0.2 mile to offices of U.S.D.A. in the modified old **greathouse of Estate Bethlehem Old Works**. At 0.1 mile, a road to the left or E goes to a vacant building constructed on the foundations of the manager's house, overlooking the village ruins (long houses) of Bethlehem New Works in the gut below. This arrangement shows the classical siting of the three basic components of plantation buildings. The greathouse was usually situated at some distance north or east (upwind) of the factory and village, which could produce a considerable smell when sugar was being refined. The manager's house was usually close to the village, uphill for surveillance purposes, and upwind. The village was often right by the mill and factory.

E-W (W-E)

<0.25>

0.5 (5.25) **Ruins of village of Estate Bethlehem New Works** (long houses where slaves lived) to right or W. The ruins are typical of the era of sugar plantation construction when the slaves' quarters were long buildings with an apartment arrangement. This economized on labor and time. Later slave quarters were even longer [see Hogensborg, p. 37, mile 12.2(3.2)]. Good examples of the early layout of individual buildings are at Estates Slob [p. 36, mile 7.4(8.8)] and Negro Bay [p. 33, mile 0.6 off Centerline at 9.8(5.6)].

<0.5>

1.0 (4.75) Paved road to right or W to Golden Grove **Territorial Prison**. The road continues between two pillars which were constructed to mark the entrance to a World War II Army Air Base (now the airport).

<0.25>

1.25 (4.5) Junction at traffic light with Four-lane Highway, Route 66 [p. 51, mile 3.2(3.0)]

<0.2>

1.45 (4.3) Junction. A left or E turn leads to **Krause Lagoon, a mangrove.**

Road to Krause Lagoon:

- 0.0 Junction with Route 64 at mile 1.45(4.3)
- 0.3 **Ruins and chimney of Estate Anguilla** on a hill to the left as road veers right or S.
- 0.9 Slaughterhouse on right
- 0.95 Public dump on right. Despite what you might see elsewhere, this is the only official dump on St. Croix. The road changes to dirt here.
- 1.15 End of road at **Krause Lagoon mangrove**. To get into the lagoon you must cross on trash bridges over the mud. There are many dikes (constructed for effluent ponds of the old alumina refinery) to walk within the lagoon for **excellent birdwatching**.

<0.15>

1.6 (4.15) On left or E, rubble of **mill of Estate Bethlehem Middle Works**. Without care, mills die.

<0.05>

1.65 (4.1) Bridge over Fairplain Creek

<0.25)>

1.9 (3.85) Two dirt roads to left or S are easy to miss (slow down!) They are on the western part of the main curve of the paved road. The left or eastern dirt road (four-wheel drive) goes 0.2 mile to the **mangrove forest of Fairplain Creek**. You will have to wend through brush for the last 50 yards. One can walk through mangrove forest along the bank of the inlet to the ocean.

-The right or western road leads to an **excellent birdwatching area** of the Manning Bay Lagoon **mangrove**.

E-W (W-E)

The road is rough, and you will have to park and walk before the road ends if using a two-wheel drive vehicle.

Manning Bay Lagoon:
0.0 Junction with Route 64 at mile 1.9(3.85)
0.3 Road divides; Take the right (westernmost) road which runs inland among **Manchineel Trees**. This entire section of road is a good birdwatching area.
0.6 Road ends. Continue on footpaths to a forest of **large Red Mangroves** with arching aerial roots extending into the lagoon.

<0.45>
2.35- (3.4-
2.55 3.2) Alexander Hamilton **International Airport**, (entrance - exit). You may visit the **air traffic control tower** M-F 8am-4:30pm. Go into the first hallway east of the baggage claim area, then go up the stairway to the tower door. Use the call button to say you are a visitor, then pull the door open while you hear the buzzing that unlocks it.

<0.1>
2.45 (3.3) Miniature mill and stone house on left or S were recently constructed. Certainly ruins of mills and stonework construction in pastoral settings are the very essence of St. Croix. (You can hardly find a page in this book that lacks a mention of them.) So it seems appropriate to have stone buildings as a welcoming landmark. However, the ruins of Estate Manning Bay were scraped away to make the airport. As recently as 1985 some remaining plaster pillars and an entryway were removed from the racetrack. It seems pathetic to have to use a model as a welcome. One hopes that there will someday be legislation passed to invoke preservation of historic ruins as an obligation for all new construction. For the time being then, enjoy the true ruins you see, for their future is insecure.

<0.45>
2.9 (2.85) Dirt road to left or S opposite the Esso yard. This road goes 0.2 mile south and provides access to the **mangroves** of Manning Bay Lagoon. The road ends in the bordering **Manchineel Trees**. Walk E to reach the mangroves; S through the Manchineel forest leads to the beach. See those nice green apples all over the ground? A lot of people have died finding out they taste fine. The Indians poisoned their arrows with them.

<0.35>
3.25 (2.5) Dirt road to left or S goes 0.35 mile ending at shore. From here one can park and walk about 0.15 mile to the western side of the mangroves of Manning Bay Lagoon.

<0.75>
4.0 (1.75) Dirt road to left or S goes 0.3 mile to private house among **ruins and mill of Estate Betty's Hope**

<0.4>

E-W (W-E)

4.4 (1.35) **Baby house** in front of Texaco tanks to left or S of road. The slave babies were left in the shelter of houses like this while the mothers worked in the cane fields.

<0.3>

4.7 (1.05) Junction with Four-lane Highway [p. 51, mile 6.2(0.0)].
-Just past the junction on the NW corner is a dirt road to left or W running 0.3 mile to **ruins of Estate Enfield Green** which has both a **large and small mill**. The large one had a double floor and is a low structure. The smaller one is typical of the mills built to operate water pumps. The ruins of the greathouse are covered in brush nearby.

<0.4>

5.1 (0.65) Junction with a dirt road. A left or W turn goes 0.5 mile to a private school built around the **mill of Estate Old Pye**. **Ruins** of a few other structures are on the school grounds. It is speculated that the oldest sugar works and plantation on the island were in this area, built by Dutch and/or English in the early 1600s, long before St. Croix was carved up into estates by the Danish. The mill is possibly the oldest mill on St. Croix. It was originally a "whim" or animal mill, with the first floor constructed around a center post of stone that arched out to become a vaulted ceiling. The cane rollers were above on the second floor. Another example of this older construction is the mill at Estate Two Brothers [p. 62, mile 0.1(1.5)].
-A right or E turn goes 0.1 mile to Cruzan Rum Distillery built among the **mill, chimney and other ruins of Estate Diamond. Tours of the Distillery** are given M-F, 8:30-11am and 1-4:15pm. Telephone for the visitor's pavilion is 772-0799. This distillery made 3 million gallons of rum in 1983. Only 5% of the production is bottled locally; 95% is sent to New York and bottled there. Distillation is only done about half the days of the year.
-The trees with very spiny branches (a pair of spines at every leaf node) lining the south side of the entry road are **Catsclaw** or Bread-and-cheese This is a native tree of dry coastal forests of Tropical America and the West Indies. The terminal branch ends are often undulating but little divided, giving the outline a spindly look. The pulp around the seeds is edible, but not really tasty.

<0.65>

5.75 (0.0) Western junction of Airport Road with Centerline [p. 36, mile 11.1(4.3)]

SIDE TRIP TO SANDY POINT

Sandy Point is a spectacular three-mile stretch of white sand beach, a regular nesting area of the enormous Leatherback Sea Turtles. The beach is also used for nesting by other sea turtles, the Green and the Hawksbill.

The road passes by a fine birdwatching area, the southern end of Westend Saltpond.

E-W:
0.0 Junction of Route 661 with Centerline (Route 70) [p. 40, mile 14.4(1.0)]. On the SW corner behind or W of the buildings, surrounded by tires, is a **Mespel Tree** (also called Sapodilla and Chicle). The sap of this species is the source of latex for chewing gum. The tree has shiny, emerald green pointed leaves, and round, brown edible fruits with sweet tan flesh and milky sap. Native to southern Mexico and Central America, they have been distributed throughout the tropics for fruit and shade.
<0.6>
0.6 Turn right or W onto dirt road. The Four-lane Highway is being extended to this point.
<0.05>
0.65 Junction with dirt road to left or S. This road leads 0.15 mile to the beach at **ruins of Estate White's Bay**. White's Bay has an emergent shelf of beach rock (cemented corals and sand) and the shoreline rocks to the W are *diente del perro* ("dog teeth") formed by water erosion. Numerous **Rock-boring Sea Urchins** have burrows in the submerged rock. This area of the island has white sand beaches, but the water is usually murky, possibly due to dredging operations at the mid-island industrial complex. The brown streak often seen about 0.5 mile offshore is effluent from the Cruzan Rum Distillery.
<0.5>
1.15 Dirt road to S. Stay right to continue to Sandy Point. A left or S turn goes 0.15 mile to end on a beach access.
<0.5>
1.65 Left or S sand road to beach, at 0.05 mile. You are now in the **Sandy Point National Wildlife Refuge**, created for the Leatherback Turtles, but encompassing a wonderful assortment of other creatures and plants.
<0.1>
1.75 Southeastern end of Westend Saltpond. Many shore birds use this pond, including nesting **Black-necked Stilts** and **Least Terns**. A road to left or S leads to beach access.
<0.3>
2.05 Salt flats and ponds to left or S of road
<0.2>
2.25 Track to S to beach
<0.05>
2.3 Southwestern end of Westend Saltpond
<0.05>
2.35 Road (no vehicles) to N along SW end of the saltpond provides access for afternoon and evening birdwatching.
<0.05>
2.4 Sand mining operations to right or N created the ponds here, now used by nesting **Least Terns** and other shorebirds. When the National Wildlife Refuge was created, some earthmoving for the preserve was done by the mining company in retribution for various

E-W:

violations. They built the berms here that protect the ponds from vehicle traffic.
-Continue straight or W, and drive slowly around the margins of pits in the road.
-A sandy road near the sand-mining pools goes left or S to the beach.

<0.4>
2.8 Access to south shore. Park and walk less than 0.05 mile.
<0.05>
2.85 Red tower of Southwest Cape Light

Fig. 20. A Leatherback Turtle nesting on Sandy Point. Photo by Scott A. Eckert.

<0.1>
2.95 Fork. A left or S turn goes to parking area for southern end of Sandy Point Beach. (Sandy Point is the point to your right as you face the water of Broad Bay. To your left is Southwest Point.) This is the most important location in the United States for **regular nesting of the Leatherback Sea Turtle**, the largest living turtle species (over 6 feet long, some over 1000 pounds). From March through June, these immense reptiles beach themselves and plow ponderously out of the surf. The turtle digs a deep pit in the sand and deposits about 80 eggs, larger than golf balls. She burys the eggs, then with oar-like flippers, flings sand in all directions, erasing traces of the nest. She hauls her half ton back to the sea, only to repeat this nesting imperative several more times in the

same season. Hatchlings emerge from the nests from May through August, after being incubated by the sun for two months. After sundown, the cooling sand triggers the baby turtles; they tunnel upward, break the surface of the sand where a bikini may have lolled only hours before, and instinctively head for the sea. Virtually all of this spectacular activity begins after sunset and continues until dawn. If you like nature specials on television, and you are ready for live action--this is it! Earthwatch often has a monitoring program and welcomes visitors. For information call the Bureau of Fish and Wildlife in Frederiksted, 772-1955. The Bureau can usually give you contact information for the Earthwatch group.

-A warning--all sea turtles are protected by Federal law. If you buy or collect any sea turtle product, such as a turtle shell or stuffed baby turtle, it will be confiscated by U.S. Customs without compensation to you. You could also be fined.

-The **endangered Least Tern** sometimes nests on this beach during the spring and summer.

-This beach is one of the longest stretches of sand on the island. It is an area of surf and undertows and favored by **surf fishermen**. Some **shells** that are common here are rare on the rest of the island.

-A right or N turn at this fork leads to other access points for Sandy Point Beach, with parking at 0.2, 0.3, 0.35, 0.4, and 0.45 miles. All parking is within less than 0.05 mile of the beach. To the inland side of this northern fork is the **unique vegetation and habitat** of the Sandy Point area. **Birdwatchers** can see Caribbean Elaenia, White-crowned Pigeons, Yellow Warblers, and numerous shorebirds. The vegetation has scrub acacia, Seagrapes, and beautiful, ground-dwelling, tall stemmed orchids in shades of lavender, brown, and ochre. It has no general common name and is locally called **"Sandy Pt. Orchid"**, although it occurs at other places on St. Croix. This orchid is common all over the West Indies and often has acquired a local name that describes where it grows abundantly on the island. It typically seems to find conditions just right in particular locations (usually dry, with dense scrub), and then grows luxuriantly in those restricted areas, oftentimes just one place on small islands. For example, on St. John it is called "Chocolate Hole Orchid" because that is one area where it is quite abundant. Perhaps it should be called the "Band-wagon Orchid", because of its colors and because of the way individuals proclaim the glory of the best site. Please leave them for all to enjoy.

TOUR OF FREDERIKSTED
and Connecting Link to Route 63

Frederiksted is currently the quieter of St. Croix's two towns. Historically it has experienced a good bit of turmoil. Because much of old Frederiksted was burned during the labor riot of 1878, many of the present buildings were constructed during the Victorian era of the late 1800s. Typically they are embellished with the gingerbread woodwork popular during that period. This tour includes the major features of the town, and then goes north to join Route 63, a scenic drive along the west shore.

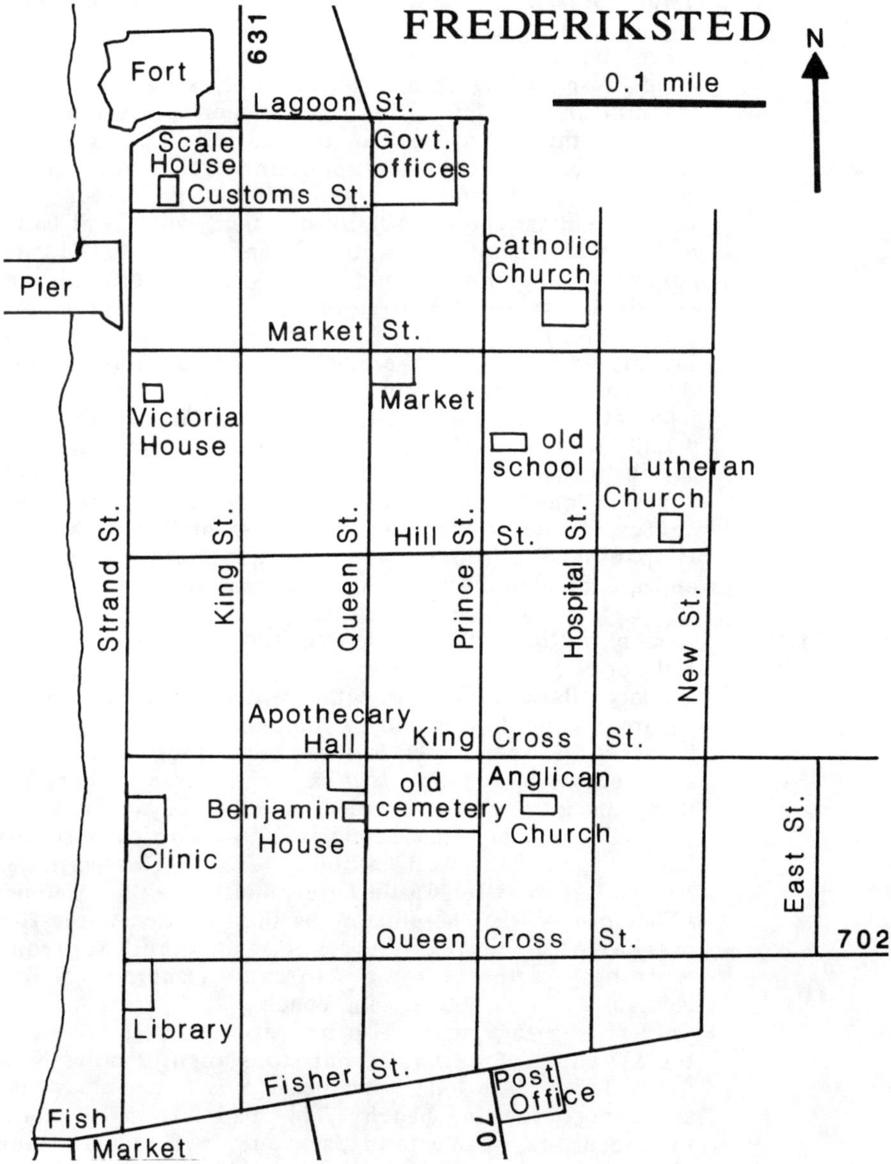

Fig. 21. Map of Frederiksted

S-N (N-S)

0.0 (1.6) Junction of Route 70 (Centerline) [p. 41, mile 15.4(0.0)] with Route 702. A **Post Office** is on the SE corner. From Route 70, go left or W here onto Route 702 (Fisher Street).

S-N (N-S)
<0.1>
0.1 (1.5) A paved road to left or S goes past **ruins and mill of Estate Two Brothers**. The mill is on the right or W at 0.1 mile. A dirt road to right or W leads to the mill's seaward side. When you are just seaward of the mill, you will see a path amid brush leading up a break in an old stone wall and to the mill on your left. Just before entering the mill, one can peer through a break in the old red brick floor into the mill basement. The supporting arches and central column of the basement story are all well-preserved. This was the oldest type of mill construction, which was based on the structure of the "whim" or animal mill. The cane rollers were on the second floor. A wooden staircase inside leads to the mill top, which is used as an observation tower by the Atlantic Fleet Weapons Training Facility. (See p. 73, mile 3.9(3.05).] Two **Shortleaf Figs** (also called "Strangler Fig" and "White Fig") are working hard at reducing the mill to rubble. [See p. 75, mile 4.75(2.2) for more about Strangler Figs.] The mill also has a rich growth of a **night-blooming cereus cactus**, which hangs in long cords, especially from the south side. This cactus was introduced from the islands of Cuba and Hispaniola. The private house just south of the mill appears to be an old house of the estate.

<0.1>
0.2 (1.4) Junction of Route 702 at mile 0.65 and Strand Street. Turn right or N onto Strand Street.
-A large **Banyan Tree** is on the corner. Banyan trees are fig trees which have auxillary roots that originate from the branches and grow down to the ground, root, then thicken and become supporting trunks for the tree. There are many species of fig trees which have the banyan habit, and you will see many fine examples of this particular species planted around the west end of St. Croix. This particular banyan is a **roost for Cattle Egrets and Little Blue Herons**.
-The shed on the shoreline W of this junction is the **fish market** of Frederiksted. Wednesday and Saturday mornings are the best times to catch fishermen selling colorful reef fish, lobster, and conch.
-A left or S turn at the fish market shed takes you along the SW shore of St. Croix, but stops about 2 miles N of Sandy Point. The long, sand beach S is a marvelous hike and a good **shelling beach**. The road terminates at the northern end of Westend Saltpond, a fine **afternoon birdwatching area**, with the sun behind you.

Road from Frederiksted along the SW coast (N-S):
0.0- 0.2 On the left or E the road passes by a government housing project with some **huge Banyan Trees**.
0.55 Northern end of **Westend Saltpond**. From this point S are good vantages for **afternoon birdwatching.** A

S-N (N-S)

Road from Frederiksted along SW coast, N-S (continued): dirt road to left or E goes around the NE end, providing sites for morning birdwatching.

0.65 **Access to beach** on right or W at mileages 0.65, 0.75, 0.9, 1.0, 1.05, 1.1, 1.2. Various common strand trees (Otaheita, Seagrape, Tamarind, Acacia) provide shade at the vegetation line. The inland side bordering the saltpond is a pure stand of **White Mangrove Trees** from mileages 1.15-1.35.

1.35 End of road at Vincent Mason Coral Resort (also called Stony Ground), a **public saltwater swimming pool, beach, picnic area, and playground**. The walk to Sandy Point is about 2 miles south from here.

Fig. 22. Getting the best price at the Frederiksted fish market.

<0.05>

0.25 (1.35) In the N end of the first block from Fisher Street is the **public library**, open M-F 8am-5pm. The stairs have bell and star emblems, placed there by the Bell family who lived in this house in the late 1800s.

-Directly S of the library is an open air theater where a **steel (drum) orchestra** practices, and the public is welcome to listen. They practice the entire school year, M-Th 7-9pm with some short, irregular exceptions. It is always amazing that such lovely muted music comes from something as mundane as a hammered oil barrel.

S-N (N-S)

-In the parking area in front of the N entrance of the library are two large trees with gray bark just to the left of the driveway. These are **Genip Trees**, a popular summer-bearing fruit tree, naturalized from Central America. The N tree in the dirt is female, the southern tree in the pavement is a male. The fruits have a large seed encased in a thin delicious pulp. Be careful, the "colorless" juice stains clothes permanently brown.

-The Genip Trees and a group of large trees to the E of the library often have roosting **Cattle Egrets and Little Blue Herons**. Large inland trees in the next block north are used for nesting.

-The tree in front of the library with lacy leaves and stalks of mildly fragrant, showy yellow flowers in spring through summer is the **Yellow Flamboyant**. This tree has been introduced worldwide in the tropics from Southeast Asia. Young branches are covered with rusty hairs. The seed pods are covered with rusty hairs initially, and hang on the tree for most of the year.

<0.05>
0.3 (1.3) In the second block of Strand Street on the right or E is Ingeborg Nesbitt Clinic, the **public clinic** for Frederiksted. Directly across the street next to a wall made of rock and coral is a **Lignumvitae Tree**, fighting it out with the broken pavement of a parking lot.

-The entire seaside border of Strand Street on your left is the **strand park** of Frederiksted, a pleasant area for a picnic and watching seabirds.

<0.2>
0.5 (1.1) Near the southern corner of the fourth block is Richards & Ayer Real Estate Office where **a wonderful collection of artifacts obtained under the Frederiksted Pier** (and other places) are on display.

-Next door is Cruzan Divers **Dive Shop**. Telephone 772-3701, open every day.

<0.05>
0.55 (1.05) On the right or E towards the north end of the same block with the dive shop is **Victoria House**, a private home with its gingerbread well preserved.

<0.05>
0.6 (1.0) On the left or W, stretching about 1800 feet out into the water, is the **Frederiksted Pier**. Park somewhere along the strand park before you get to the foot of the pier and walk out onto it. (If you are taking diving gear, drive onto the pier since it is usually open to vehicle traffic.) There are usually **handline fishermen**, and sometimes one with a **cast net getting bait fish** for deeper water fishing.

-**Birdwatching at the pier** is always enjoyable. One often sees Brown Pelicans, Brown Boobies, Magnificent Frigatebirds, Terns, Laughing Gulls, and Caribbean Martins.

Frederiksted Pier (continued)
-This pier is a famous **snorkel and dive site** of the Caribbean. Depths beneath the pier range from 10-40 feet, averaging 25-30 feet beneath the fabulous wide section. If military vessels or cruise ships are at the pier you need to request permission from Port Authority to dive. (Their building is just to the left or S of the main pier gate.) Entry is easiest by taking a long stride out from the pier edge and dropping to the water 6 feet below, or one can use the stainless steel ladder at the E end of the wide section of the pier. Also one can enter from the shore end of the pier by walking in over the beach rock. One place to enter over the riprap is at the seaward end of the small boat docking area on the S. If a long snorkel does not worry you (and if you snorkel along over the riprap, it is interesting anyway), you can also enter by walking in over the beach rock behind the clock tower, or at the public beach at the north end of Frederiksted--see mile 1.5(0.1) below. Exits are made at the ladder on the broad section of the pier or the riprap. If climbing up a ladder with a tank or other equipment like cameras is a problem, take along rope to tie to one of the bollards on the top of the pier and dangle the end into the water near the ladder. Then at the finish of the dive, tie your equipment to the line, climb out, and haul the equipment up afterwards.
-Because of the broad end section of the pier, the interior pilings are always shaded. Due to this low light level, they are festooned with **sponges** and other fauna that are found naturally in abundance only at greater depths where light is diminished. **Fishes are very diverse.** One often sees fishes that are rarely seen elsewhere: such as Seahorses clinging with their tails to the sponges; Batfish walking around the bottom on their peculiar fins; Frogfish, exasperatingly camouflaged; and foraging Flying Gurnards flipping pebbles around with their fins. There is usually a colony of Garden Eels on the sand just west of the junction of the cat-walk and the broad section of the pier. Collectors of **old bottles and artifacts** like to dive the pier after storms or after cruise ships have docked because the roiling of the bottom often brings up long-lost jetsam. For deep SCUBA divers, one can snorkel out about the distance of the catwalk again beyond the pier and be in about 90 feet of water. To make this dive you should snorkel until you begin to lose sight of the bottom, and descend while you can still see it to about 70 feet under normal visibility conditions. There are scattered colonies of **black coral** on the deep reef here. A night dive under the pier is a spectacular experience, with a cast of nocturnal characters quite different from those of the day.
-Across the street from the pier is the **Old Scale and Customs House** with the scale set up so that one can see its

Frederiksted Pier (continued)
inner works. The **Visitor's Bureau** is on the ground level and has pamphlets describing walking tours of Frederiksted and Christiansted. Maps and other information are also available.
-Just north of the pier is a clock tower (which confirms the promotional that told you this is a place where time stands still). **Public rest rooms** (not always open) are in the base. North of the clock tower through the park with the bandstand is the entrance to **Fort Frederik**, constructed from 1740-1760. **Museum** hours are M-F, 8:30am-4:30pm. The fort has **rotating historical and archaeological exhibits**. It also has **public rest rooms**. Two notable historical events occurred at the Fort. In 1776 this fort gave the first foreign salute to the flag of the new American Republic. In 1848 Governor General Peter von Scholten signed a proclamation here which emancipated the slaves of the Danish West Indies.
-The trees in the bandstand park are **West Indian Mahogany**.

S-N (N-S)
0.6 (1.0) Turn right or E at the gate of the pier so that you travel on the S side of the Old Scale House along Customs Street. The first cross street is the main street of Frederiksted, King Street or Route 631. Cross carefully.

Fig. 23. Frederiksted market. Buy a drink of maubi or ginger beer.

S-N (N-S)
<0.1>
0.7 (0.9) On the right or S in this block is a **huge Kapok Tree**. The massive buttress roots are typical of large trees of the tropics. In summer you may see the floss of the seedpods, which suspends the seeds so that they can float some distance in the wind. This floss is kapok, used extensively for stuffing before the advent of synthetics.
<0.05>
0.75 (0.85) Turn right or S on the next street, Queen Street.
<0.05>
0.8 (0.8) Turn left or E onto the next street, Market Street. At this intersection on the SE corner is the **open-air market** of Frederiksted. This market dates from 1751 when the town was laid out. As in Christiansted, there are often vendors here selling local produce. Try a fruit or drink you've never seen before.

Fig. 24. Tropical fruits from left to right: Mango, Soursop, Breadfruit, Sugar-apple. Buy them at open-air markets and stands along the roads.

<0.05>
0.85 (0.75) Turn right or S onto the next street, Prince Street. On the NE corner of this intersection is St. Patrick's Church, which was dedicated in 1844. The church yard has several very large **West Indian Mahogany Trees**.
-Just inside the front gate of the cemetery is a monument to several sailors. A major earthquake hit St. Croix in 1867 and the resulting tsunami (tidal wave) tossed the *U.S.S. Monongahela* from the Frederiksted harbor onto Strand Street. Some of the sailors who drowned were buried at St. Patrick's; others were buried at the Holy Trinity Lutheran Church Cemetery, also with a monument. The Lutheran cemetery is on the NE corner of Hill and Hospital Streets. Both cemeteries have interesting gravestones. The *Monongahela* had brought a group of U.S. officials on a junket to determine if the U.S. should

S-N (N-S)

purchase St. Croix. A treaty for the sale of St. Thomas and St. John had been signed and was awaiting ratification by the U.S. Senate. Denmark was hoping to fit St. Croix into the sale also. The tsunami struck while the officials were safe on the land, living it up as the recipients of enthusiastic lobbyists. Despite all the free trips, the Senate did not ratify the sale: too expensive ($24/acre compared with two cents an acre just paid for Alaska), too many hurricanes and earthquakes, and "Why do we need a naval base in the Caribbean anyway?" Meanwhile, the *Monongahela* graced Strand St. for a year before it could be maneuvered back to sea.

<0.05>
0.9 (0.7) In the middle of the next block on the left or E is an **old Danish school**, now the Dept. of Social Welfare.

<0.05>
0.95 (0.65) In the yard just north of a large (currently burnt-out) house on the NE Corner of Prince and Hill Streets, is a large **Flamboyant Tree** with a trunk diameter of about 1 foot. Long-lasting, large flowers in shades from red to yellow cover these trees in spring and summer, making them a favorite ornamental. They have lacy leaves and black woody pods over a foot long that can be found on the trees all year. Originally from Madagascar, they have been introduced world-wide in the tropics.

<0.1>
1.05 (0.55) One block from the corner of Hill and Prince Streets, turn right or W onto King Cross Street. [Park near the corner to explore this intersection and the one described below at 1.1(0.5).] On the SE corner is St. Paul's Episcopal Church built in 1812. In front of this church the sidewalk deviates around a nice large **Tamarind Tree**. This species has been planted worldwide in the tropics. In summer the pods are harvested and the tart pulp around the seeds is used to flavor many kinds of dishes and drinks. Grocery stores in the states have recently started selling tamarind pods.
-Directly W across the street from St. Paul's Church is the old **Frederiksted Cemetery**, which is no longer used. Beneath large Tamarind and West Indian Mahogany Trees are old stones and monuments, most dating from the late 1700s and early 1800s.

<0.05>
1.1 (0.5) Turn right or N onto Queen Street. From here you will be going 4 blocks straight or N to Lagoon Street, but first look at the buildings on this corner.
-Directly west across Queen Street from the cemetery are two old buildings of interest. The building on the SW corner of King Cross and Queen Streets with the arcade around its ground floor is called **Apothecary Hall** and is extensively ornamented with the gingerbread typical of the

S-N (N-S)

post-fire architecture. The private residence just south of Apothecary Hall at 48 Queen Street is the **Benjamin House**, one of a very few buildings that escaped the 1878 labor riot fires. It was owned by a judge whom the workers liked. It is made chiefly of brick, has a wrought iron balcony on the front of its second story, and is graced by a varied arcade on the ground floor.

<0.3>

1.4 (0.2) Queen Street and Lagoon Street. To continue, you will be veering slightly left or NW on Queen Street, passing beside the Children's Playground. The complex of government offices on the SE corner of this intersection has the **Bureau of Fish and Wildlife Office** for St. Croix, telephone 772-1955, Room 203, upstairs in the eastern building. On the ground floor of the NW building is the **Office of Archaeological Services**, with some exhibits of artifacts exhumed on St. Croix. Their telephone is 772-2021, 772-2250.

-The **trees** around the Children's Playground are easy to point out for **identification**. Going north on Queen Street: a Kapok Tree; (gate); Puerto Rican Royal Palm (also native to St. Croix); Kapok; Kapok; West Indian Mahogany near the junction with Route 631. If you were to make a 160° turn onto Route 631 back towards town, you will pass seven Horsetail Trees along the fenceline. Near the corner of Lagoon Street and Route 631 is a small native tree, Pasture Fiddlewood planted for ornament because it bears small fragrant white flowers and has attractive orange bases on its leaves. Its very dense wood has been used to make musical instruments, hence the name. Now turn left or E to complete the circle around the playground. The remaining trees are a Horsetail Tree and Tamarind. Along the fenceline the low plants with straplike leaves are native **West Indian Spider Lilies**, which grow wild on St. Croix and most of the West Indies.

<0.1>

1.5 (0.1) Junction with Route 631. Veer right or N onto 631 and cross a bridge over a small stream or gut. Just beyond on the seaward side is a beach house with colorful murals. This is the **Frederiksted beach**. The beach house has **fresh-water showers** and is open daily 9:30am-4:30pm. This beach has interesting **snorkeling** around large boulders scattered in the sand just offshore. You can also explore the rock below Fort Frederik.

<0.1>

1.6 (0.0) Junction with Route 63 coming in from the right or E [p. 72, mile 2.2(4.75)]. Continue straight or N to travel along the W shore on Route 63. On the seaward side just N of the junction is the private St. Croix Country Club. Just beyond is the LaGrange Beach and Tennis Club, open daily from 10am-6pm, $2 fee.

Fig. 25. Native fish traps (called "pots") are hand-made. Fairs on St. Croix display such arts and crafts.

Fig. 26. Well-kept gingerbread woodwork in Frederiksted.

WEST END ROAD, ROUTE 63

(Also Northside Road, John David Merwin Highway, Veterans' Road)

This road runs along the northern part of the west coast of St. Croix, passing along the edge of the wettest and greenest part of the island. The west end of the island averages 40 inches of rain per year, while the east end averages only 20 inches. Most of the beach has public access and provides good snorkeling for rocky outcrop and grass bed habitats. Three major roads of tourist interest intersect the West End Road: Mahogany Road (Route 76), Creque Dam Road (Route 58), and the Scenic Drive (Route 78). John David Merwin was the V.I. governor responsible for construction of the Frederiksted Pier and the Scenic Drive; this road is named in his honor.

S-N (N-S)

0.0 (6.95) Junction with Centerline (Route 70) [p. 40, mile 13.6(1.8)]. The first 2.2 miles of this road bypass Frederiksted.

<0.05>

0.05 (6.9) A dirt road on the right or E leads to private **ruins and mill of Estate Concordia**. An old coral block wall of this estate is on the right or E side of the road.

<0.15>

0.2 (6.75) The schoolyard and entry road on the N have **Tibet Trees** (also called Old-woman's-tongue and Shek-shek after the clattering sound made by the numerous seed pods). This tree from tropical Asia has been widely planted on the western half of St. Croix. The characteristic seed pods are green to rusty when recently formed and dry to a straw color. During the period of dry seed pods and when Tibets drop their leaves, they are easy to distinguish, and it becomes apparent what a common tree it is.

<0.80>

1.0 (5.95) Junction with Route 702. Veer left or W, then right or N 0.05 mile further to continue on Route 63. Route 702 to W goes 0.5 mile into Frederiksted. Dirt road to right or E provides a **view of Frederiksted**.

Route 702 to E uphill with view of Frederiksted (W-E):

0.0 Junction with Route 63 at 1.0(5.95) and Route 702
0.3 Fork. Go left.
0.4 Three-way fork. Take the middle road on pavement.
0.55 Pavement ends.
0.65 Dirt road to left or N. About 100 feet north of the junction, this offshoot provides a fine **view of the Frederiksted area** from Prosperity Marsh on the N to the pier and then to Sandy Point on the S.

<0.3>

1.3 (5.65) Paved road to right or E. In the field north of this road are many large trees with lacy leaves borne on spreading crowns. These are **Raintrees** (or Samán), native from Mexico to Brazil. From spring to fall they have pink, powder-puff flowers.

<0.4>

S-N (N-S)

1.7 (5.25) Veer left or W on the paved road to continue on Route 63. The private dirt road to right or E is lined with **Flamboyant Trees.**

<0.1>

1.8 (5.15) Brugal Rum Distillery among **ruins, chimney, and mill of Estate LaGrange** on the right or N. At the end of the 19th century LaGrange was converted to a steammill and became a central factory for West End sugar estates. Note the brick chimney; a rarity in that most were made from coral block. The distillery, operating today, replaced the sugar factory.

<0.4>

2.2 (4.75) Junction with Route 631 [p. 69, mile 1.6(0.0)] on the coast. Turn right or N to continue Route 63. The private St. Croix Country Club is just seaward and N of the junction. Next north is LaGrange Beach and Tennis Club. Day use fee is $2, open daily 10am-6pm.

<0.2>

2.4 (4.55) Junction with Mahogany Road, Route 76, on the inland side. This is a scenic drive on paved roads through part of St. Croix's **"rain forest"**. [See "Side Trip on Mahogany Road", p. 81, mile 0.0(9.65).]

<0.3>

2.7 (4.25) Sunset Beach Cottages on the seaward side and road inland into private residential area amid **chimney, factory ruins, and village of Estate Prosperity.**

<0.25>

2.95 (4.0) Dirt road inland. At 0.05-0.15 mile one can view **Prosperity Marsh** over the left or N fence. This is one of the few year-round marshes on St. Croix. In the winter, the mangroves bordering the marsh on the north are a good place to see **migrant warblers.**

<0.05>

3.0 (3.95) Seaward is the Rainbow Beach restaurant. Inland is Prosperity Marsh. The **mill of Estate Punch** is on the hill to right or E.

<0.3>

3.3 (3.65) Private house built on **ruins and mill of Estate William** on the inland side.

<0.1>

3.4 (3.55) **Ruins of slave quarters of Estate William** inland

<0.15>

3.55 (3.4) Sprat Hall Beach and Restaurant, seaward

<0.15>

3.7 (3.25) Junction with Creque Dam Road (Route 58). This road winds inland through a **scenic water course** and St. Croix's **"rain forest"**. It joins the Western Scenic Road 3.1 miles away from this junction. [See "Side Trip Along Creque Dam Road," p. 88.]

-In front of the hotel cottages on the NE corner at this junction is a **Kapok Tree**, distinguished here by its large buttress roots. The other large trees are **Raintrees.**

S-N (N-S)
<0.05>
3.75 (3.2) Entrance to Sprat Hall Plantation on inland side. The main building of this hotel is the **greathouse, furnished with antiques**. There are **extensive cleared ruins** on the grounds and a **mill with a scenic view of the west coast towards Frederiksted**. There are many kinds of **horticultural and native plantings, some of them labeled**. The only **guided horseback trail rides** on St. Croix leave from the hotel, usually at 10am and 3pm every day. Rides are through scenic areas of tropical forest and historical ruins. Moonlight rides are available on and near the night of full moon. Call for reservations, Jill's Equestrian Stable, telephone 772-2880, 772-0305.

Fig. 27. Large iron pots (called coppers) like this one at Sprat Hall Plantation were used to boil down juice squeezed from sugar cane.

<0.15>
3.9 (3.05) Entrance on seaward side to **Atlantic Fleet Weapons Training Facility**. From the shore beyond these buildings, cables run to connect with a huge grid of underwater receivers. The system extends to a depth of about 10000 feet and 10 miles offshore. The U.S. Navy uses the system for testing and training with many kinds of weapons, particularly torpedoes and missiles. For this reason there are often submarines at the Frederiksted Pier. This site was selected for its quiet waters, which have a low volume of ship traffic, and in fact, it is the quietest of five training sites like it.

<0.2>

S-N (N-S)

4.1 (2.85) Dirt road inland passing through a shady forest. You may drive the road to view the nicely cleared historical sites of **Estate Mount Washington**, but the owners request that you stay outside the fence line. You may also **hike to the top of Mount Washington**, where the **view of Frederiksted, the pier, and Sandy Point beyond** is superb.

Estate Mount Washington Road (W-E):

0.0 Junction with Route 63 at mile 3.75(3.2)
0.35 Old **village cemetery** on the left
0.8 **Ruins of sugar factory and animal mill** on the left
0.9 Three-way fork. Take the middle road to continue. The next 0.3 mile is usually passable by two-wheel drive, though steep and rocky.
1.15 Three-way fork at a **viewpoint of Sandy Point and the Frederiksted Pier**. To continue to the top of Mount Washington (also called Frenchman Hill, 807 feet), hike the ridgeline up the middle road or trail (foot traffic only). The trail passes through moist forest for about half the distance, about 0.4 mile to the top. The top of **Mount Washington affords a superb view of the plain below stretching to Frederiksted and beyond to the southernmost tip of the island, Southwest Cape.** To the NE one can see the **mill of Estate Nicholas** with a great factory ruin at its base, the **mill of Estate Mount Victory** on a hilltop nearby, the **mill of Bodkin** on the ridgeline to the east, and the **mill of Punch** on a hill to the SE.

<0.05>

4.15 (2.8) Road makes a sharp "S" over Great Gut. The bridge has plastered rock walls. This is a landmark for the snorkel at the next stop.

<0.25>

4.4 (2.55) Turnout for **snorkeling** in Butler Bay area. The beach is largely coral rubble, the bottom offshore is a nice seagrass bed. If you are willing to do a long snorkel, just S of Great Gut in about 8 feet of water you will find boulders with deeply pocketed surfaces and many small caves.

<0.1>

4.5 (2.45) Large-trunked, peculiar tree on right or E is a **Baobab Tree**, introduced from Africa. Directly across the road from this tree is an old **lime kiln**, an oven used to cook coral to make mortar.

-About 25 yards further N is a paved road entrance to Estate Butler Bay (also called Prospect Hill). Turn in between the plastered stone pillars to see **large Baobab Trees** growing among **ruins of Estate Butler Bay** on the right. According to one theory, the planters introduced them as a curiosity, but according to another theory, the slaves brought the Baobab with them because it is a tree of many uses. In Africa they are used as shelter, and both

S-N (N-S)

the fruits and the leaves provide human food as well as forage. Leaves and bark provide clothes, paper, rope and medicine. In Africa the Baobab is called the "Upside-down Tree", because the branching pattern and the trunk's outward curve look like exposed roots, especially when leaves drop. These trees have not naturalized here, and are found at only a few places on St. Croix. The one with the largest girth demonstrates how the old trees maintain such a size with so little foliage--the interior is hollow.

<0.1>

4.6 (2.35) Private **greathouse of Estate Butler Bay** on right or E. Looking back to the S, you will see the **mill** of Butler Bay, incorporated into a house on the hill above the old factory.

<0.15>

4.75 (2.2) Ruins of a small pier in the water mark a favorite beach and picnic area of Butler Bay. The sea bottom is a fine seagrass bed. The inland dirt road (15 yards N of the green mileage sign "5.5") provides access to the **gut of Butler Bay** and a delightful **hike through moist forest to an 80-foot waterfall**. (No water will be falling unless there has been recent substantial rain.)

Hike up Butler Bay Gut (W-E):
From the junction with Route 63, walk 200 yards along the main dirt road to the SE, crossing under a power line. Ten yards past the power line is a fork. Take the right fork down towards the gut and trees. Cross the stream bed, then go left for about 100 yards, following an indistinct trail that is south of the stream bed and always within 50 feet of it. You will then reach a point where the stream bed has less vegetation and is easy to enter and hike. The hike through the valley along the stream bed (gut) is shady and has little brush. It is a gradual incline starting near sea level, reaching the falls at 300 feet elevation, 0.9 mile upstream. Remnants of an old road parallel the gut here and there. The streambed passes beside many large forest trees festooned with hanging vine roots, lichens, and **Strap Fern** with its 1-3 foot, unbranched fronds.

-There are several large **strangler figs** along the stream course. Most of these are a large-leaved kind that specializes in shady areas. These strangler figs start life when a bird defecates the seed onto the branch of a tree where it can get some light. Here they sprout leaves to start growth and send out roots that drape down towards the ground. When the roots touch dirt, they continue growing into the ground, while the aerial part thickens and becomes a trunk for the fig. Meanwhile the host tree is so squeezed by these trunk-roots and so shaded by the very large leaves of the fig, that it often dies, leaving the fig

Hike up Butler Bay Gut (continued)
the master of the spot. This insidious strategy positions the fig for effective competition in a place where space is at a premium. If it had to start on the forest floor, it might never succeed in gaining the light it needs to grow or escape being eaten by animals that forage on the ground.

Fig. 28. Some large Strangler Figs grow along Butler Bay Gut.

-Large black moths (called "**Black Witch**") will flutter from beneath your feet. Large female **Golden Silk Orb Weaver Spiders** with their tiny attendant males weave webs with golden threads. Carry a short stick to remove webbing that blocks the way. There are three abundant types of **millipedes** ("gongolo") that you may see: a brick-

Hike up Butler Bay Gut (continued)
red one with its pretty grass-green legs; a gaudy one with its yellow-green bars on each segment of its brown body, and a large black one (to 5 inches). The red one is found world-wide in the tropics; the barred type is widespread from Bermuda, through the West Indies, and South America; and the black one occurs only in the West Indies. Pearly-eyed Thrasher and Bridled Quail-Dove calls are also very abundant.

Fig. 29. Butler Bay Waterfall.

-Most persons will want to return to Butler Bay at the base of the falls. Ascending the falls is dangerous (especially when wet) but can be done without climbing aids. **Strap Ferns** are numerous along the borders of the falls. Several

Hike up Butler Bay Gut (continued)
peaceful pools are right at the top of the falls. From here one can walk the gut above the falls out to Estate Nicholas. (At one point a stream joins the main gut from the right or S. Continue on the left streambed to the E.) The gut ends in pasture of Estate Nicholas. Just as you exit the forest you will see a good example on the N of the **Strangler Fig** that is more common in less shaded, dryer forests, called "Shortleaf Fig" or "White Fig". The road out to Creque Dam Road is S of the Nicholas Mill [p. 91, mile 2.0(1.2)]. You may need permission to go out that way. You can return back down the gut or walk out to the SW over Mount Washington and down the Butler Bay residential dirt road to Route 63 [p. 74, mile 4.5(2.45)], only 0.25 mile from where you started.

<0.65>
5.4 (1.55) Paths down to the beach opposite a chain-link fence lead to an area with squarish pools that seem to be cut into the beach rock. These are purported to be "baths" made by the Knights of Malta when they owned the island in 1653-1665. Hence the name of this beach, "The Baths, Malta Baths, or Monks Baths". One pool in particular is markedly rectangular, and the sides appear hewn rather than dissolved. It is likely that these are natural formations, and that the whole idea of the baths was dreamed up in this century to glamorize the area as a resort (since bathing was considered unhealthy in the 1600s.) The steps, at least, were cut in the 1930s.

Fig. 30. Natural swimming pools at Malta Baths

<0.1>
5.5 (1.45) Beresford Manor (private residence on **ruins including a watermill, chimney, and factory of Estate North Side**) on

S-N (N-S)

inland side. From the northern end of the ruins one can see the old cane-grinding windmill on top of the hill south of the ruins.

-Trails on seaward side lead to **sandy beach**. The shoreline is **beach rock**, corals and sand that have been cemented together and raised. Some of it is quite eroded by water solution and the action of **Rock-boring Urchins** and other animals. This is a good **tidepooling area** when the surge is not strong. One can easily walk the beach south to the bath pools from here.

-The area just offshore at 15-40 feet depth is a reef of beachrock with many sponges and soft corals providing shelter for numerous fish and invertebrates, a very interesting **snorkel or dive**. Beware of wave surge when entering and exiting.

<0.6>

6.1 (0.85) Road veers right or E. Parking area on seaward side for Ham Bay. This is an interesting **snorkeling area**, but usually cloudy. It is probably unique on St. Croix because of the continual flow of fresh water into the sea here from Caledonia Gut, the only year-round flowing stream on the island.

<0.05>

6.15 (0.8) Junction with Western Scenic Drive, Route 78, at point where two plaster pillars topped with plants border a road leading in from the left or N. To continue on this tour turn left or N and go between the pillars.

-To take the Western Scenic Drive go straight or E. This is the end point of that log [p. 125, mile 19.5(0.0)]. Near this junction as Route 78 starts uphill is a wonderful **hike up the perennial stream bed of Caledonia Gut**. [See p. 124, mile 19.35(0.15) for a description.]

-Just past the pillars you will cross the stream bed of Caledonia Gut. At the NW corner of the road and gut is a large tree with a trunk over three feet in diameter, a **Raintree**, so called because it closes its leaves on cloudy days (and at night--in other words, under low light levels.) The roots hanging from its branches are from *Cissus* vines that grow on the crown.

<0.05>

6.2 (0.75) Dirt road to seaward (four-wheel drive only) goes 0.1 mile to end at parking on shore of Ham Bay. A large tree with a diameter greater than 2 feet near this junction is a **Raintree**. A small concrete shed in next to the tree.

<0.05>

6.25 (0.7) Private home on right or E, built amid **ruins of Estate Ham Bay**. The **mill** is on a hill behind or E of the hotel.

<0.15>

6.4 (0.55) Dirt road seaward to beach. Nearby is a tower about 10 feet high used by AFWTF. [See p. 73, mile 3.9(3.05) above.]

<0.3>

S-N (N-S)

6.7 (0.25) Two dirt roads seaward with parking for cobble beach of Ham Bluff. One can see **spectacular, photogenic breakers** against the bluff and beach rock on most days. The beach is quite scenic with the **lighthouse** on the bluff above. Another 10 foot AFWTF tower is on this beach. The **shoreline is a consolidated fossil coral reef**, as one can easily see from the coral and shell exposed as it weathers. It is eroded to holes, pocks, and tidepools with jagged edges, a weathering form called *diente del perro* (dog's teeth). Walk carefully; it's beautiful, brittle, and unforgiving of flesh. This formation extends eastward irregularly along the base of the bluffs of St. Croix's northwest coast, and its eroded surface forms most of the tidepools on the island. This is a very nice **tidepooling area**, but stay back from the waves. Offshore is an interesting **dive and snorkel area** with large boulders and caves along the shore, especially below the bluff. Only experienced divers should enter the water, and then only when the sea is calm. Currents can be very strong and irregular, and on most days the surf is too great for entry. To enter walk E around the little bay, Christening Hole, to the bluff and wend out through the rocks and coral.

<0.25>

6.95 (0.0) End of the tour at gate of former **Coast Guard Station**. You can barely see the top of the lighthouse from this point. Park at the side and do not obstruct the gate. You may need permission to hike to the lighthouse. The following is a description of the **hike to the lighthouse** on the bluff above, which affords a **good view or the northwest coastline**.

Hike to Ham Bluff Lighthouse:

Walk seaward (left or N) outside the fence and around the compound to a dirt road which crosses Roger Gut and leads left or E uphill. This road climbs from 60 feet to 360 feet in about 0.4 mile distance (14% grade).

-On the way, you will see a sign on the right or S "Introductory Trial. **Cigar-box Cedar**. **Primavera**. 1961."

-Walk E out to the cliffs at the lighthouse clearing for the **view of the cliffs and shoreline**. Maroon Hole is the groove below with the big, surging tidepool.

-Note the extreme windshear of all the vegetation covering the upper areas of the cliffs. Most of the nearby rock is covered by **Christmas Orchids**.

-To the E one can see the bay off Davis Beach, North Star Beach, Cane Bay Beach, the point at Estate La Vallee, and Baron Bluff beyond. Mount Eagle (elevation 1165 feet) and Blue Mountain (with all the antennae and dishes, 1096 feet), together form the highest ridge on the island.

SIDE TRIP ALONG MAHOGANY ROAD, ROUTE 76 AND MIDLAND ROAD, ROUTE 72

Mahogany Road is the only paved road through the very lush secondary growth forest (once all sugar cane fields) of the western end of the island. Most people enjoy the tall trees and hanging vines. The forest is typical of the wetter end of the island, which receives about 40 inches of rain each year. A true rain forest, however, would receive at least 80 inches, and the forest floor would be shadier and have very little undergrowth. After emerging from the forest, the road changes name, first including a small section of Route 69, and then becoming Midland Road, Route 72.

Mahogany Road, Route 76
W-E (E-W)

0.0 (9.65) Junction with West End Road, Route 63 [p. 72, mile 2.4(4.55)]. **Huge West Indian Mahogany Trees** that gave the road its name continue on both sides of the road for about one mile. Many are probably well over 200 years old. (They are interspersed with other kinds of trees, so look closely at the stop ahead for two Mahoganies that are easy to point out.)

<0.3>

0.3 (9.35) Dirt road to left or N goes to **greathouse of Estate Prosperity**. Large **West Indian Mahogany Trees** flank both sides of this entrance. West Indian Mahogany Trees are characterized by their distinctive leaves, each 4-7 inches long with 4-10 shiny green, lance-shaped leaflet pairs. The brown oval fruit has five compartments that open like an umbrella when the seeds are dry.

<0.35>

0.65 (9.0) Dirt road to left or N is to private old **greathouse of Estate Prosperity Garden** on a hill.

<0.2>

0.85 (8.8) Dirt road to right or S is residential. Start of **"rain forest"**. Just beyond (E of) the dirt road is the start of the **ruins of a concrete aqueduct**. The aqueduct was constructed in the first decade of the 1900s to carry water from a reservoir near Jolly Hill to irrigate Estate LaGrange, the first irrigation project on St. Croix. It failed because the soil of the reservoir was too porous to hold sufficient water.

<0.2>

1.05 (8.6) Near the green mileage sign "1" are the end of the LaGrange Aqueduct [see mile 0.85(8.8) above] and approximate end of huge West Indian Mahogany Trees. [See mile 0.0(9.65) above.]

<0.3>

1.35 (8.3) Junction with dirt roads to right and left. The residential road to left or N leads to a **view of Frederiksted from Acacia Hill**.

Acacia Hill (S-N):
0.0 Junction with Route 76 at mile 1.35(8.3)

Side Trip Along Mahogany Road, Route 76, and Midland Road, Route 72

Mahogany Road, Route 76

W-E (E-W)

Acacia Hill, S-N (continued)

0.05 Fork. Take the right fork.

0.15 Fork. Take the left fork.

0.3 On the left side as the road curves to right or E is a **view of Frederiksted** from Estate Prosperity on the N to Sandy Point on the S.

0.45 View of **mill of Estate Little LaGrange** on a ridge to S across the valley behind you. This is easier to see as you are returning to Route 76.

0.65 View of **mill of Estate Punch** on a hill to right or N. The road you are traveling was the route to Punch in the 1700s. The road ends at private houses at 0.8 mile.

<0.1>

1.45 (8.2) Concrete ruins (sometimes inhabited) of Estate Little LaGrange on the right or N side of the road.

<0.4>

1.85 (7.8) Junction with Route 763, a dirt road to the left or N. This road is often very rough after heavy rains. The road joins Route 765 at its northern end in 1.9 miles. [See mile 3.25(6.4) below for a description of Route 765.]

Route 763, Oxford Road (S-N):

0.0 Junction with Route 76 at mile 1.85(7.8).

0.55 Fork at a green mileage sign ".5". The right fork to E continues Route 76. A sometimes inhabited ruin of Estate Oxford is on the left. The left road is four-wheel drive only and goes 0.15 mile to private **greathouse of Estate Oxford**.

1.6 Old **wooden schoolhouse**, now a private residence, on the left.

1.85 **Ruins and mill of Estate Annaly** incorporated into a private residence on the left or N of the road.

1.9 Junction with Route 765 at mile 0.9. A left or N turn here goes 0.3 mile to join Creque Dam Road [p. 92, mile 2.95(0.25)]. A right or S turn goes 0.9 mile to join Mahogany Road [p. 83, mile 3.25(6.4)].

<0.2>

2.05 (7.6) Driveway to St. Croix Life and Environmental Arts Project on right or S. LEAP is a training center for **items made with native woods.** The Mahogany Shop at the end of the road sells beautiful woodwork and is usually open every day. Telephone: 772-0421.

<0.2>

2.25 (7.4) Bridge over Mt. Stewart Gut. Just west of the bridge is a large leaning **Samán Tree** or Raintree, a native of the continent from Mexico to Brazil, which has naturalized on St. Croix.

<0.15>

Side Trip Along Mahogany Road, Route 76, and Midland Road, Route 72

Mahogany Road, Route 76

W-E (E-W)

2.4 (7.25) Entry point for **hike in Mt. Stewart Gut** as it crosses the road. A turnout for parking is on the NE corner. One can walk up the gut about 400 yards with trees towering over you fifty feet and more, past pools of fishes (mostly introduced mollies). The palms that you see in the gut are natives, **Puerto Rican Royal Palms.**

Fig. 31. Hiking up Mt. Stewart Gut off Mahogany Road

<0.5>

2.9 (6.75) Dirt road to left or N to private **greathouse of Estate Orangegrove**. The buildings on the road are part of the old estate village. The mill (not visible from the paved road) is on a hill N of the greathouse.

<0.05>

2.95 (6.7) **Ruins and chimney of Estate Orangegrove** to left or N. Four round pillars on the roadside are the former entrance.

<0.3>

3.25 (6.4) Junction with paved Route 765 to left or N.

Route 765, Orangegrove Road (S-N):

0.0 Junction with Route 76 at mile 3.25(6.4). There are many **Mango Trees** along this road, and it is often busy when fruits are ripe in June and July.

Side Trip Along Mahogany Road, Route 76, and Midland Road, Route 72

Mahogany Road, Route 76

W-E (E-W)

Orangegrove Road, S-N (continued):

0.9 Junction with Route 763 on left or W [p. 82, mile 1.9]. **Ruins and mill of Estate Annaly** are incorporated into a private residence on the NW corner of this junction. Annaly has the largest mill on the island.

1.1 Driveway on left or W into Estate Annaly. About 10 yards beyond (N of) driveway are three fig trees with braided trunks. Another fig is opposite on the left or E. This tree species, a type of **Strangler Fig**, is found from St. Croix south through the West Indies into South America.

1.2 Junction with Creque Dam Road, Route 58 [p. 92, mile 2.95(0.25)]. A left or W turn here goes down a rough dirt road through a **lush forest area** to join the West End Road, Route 63, 2.95 miles away. A right or E turn here goes 0.15 mile to join the Scenic Drive [p. 122, mile 16.15(3.35)].

<0.3>

3.55 (6.1) **Chimney of Estate Montpellier** in view to left or N

<0.1>

3.65 (6.0) Residential paved road to left or N near green mileage sign "3.5". At 0.1 mile the **chimney of Estate Montpellier** is on the left, and the foundation and stairway of a former ruin form the base of a house on the right.

<0.1>

3.75 (5.9) **Site of Chapel of Our Mother of Perpetual Help**. It burned to the ground in 1986. This chapel was of sentimental importance to islanders who were sheltered here after most of the houses in this area were severly damaged or destroyed in a 1928 hurricane.

<0.3>

4.05 (5.6) Dirt road through metal siding gate to left or N to **Estate Two Friends**. At 0.2 mile the road ends at a wire gate near the chimney and other ruins.

<0.25>

4.3 (5.35) Entry point for a **hike down into St. George Gut** on the right or S of road near a sign indicating curves ("S") in the road ahead to E. The path ends at a pool near the base of a 30-foot waterfall. In the trees above you will see broad-leaved **Strap Ferns**, while near the pools below and above the fall you can find **Swamp Ferns**, whose fronds are reminiscent of corn plants.

<0.3>

4.6 (5.05) **Mill and ruins of Estate Springfield** on hill to right or SE. At 50 yards before (W of) the turnout is a path down into **St. George Gut**. One can **hike** W up the gut for at least 0.3 mile. Near the entry point as well as along the watercourse are many **Sweet-lime Trees**. The Sweet-lime

Side Trip Along Mahogany Road, Route 76, and Midland Road, Route 72

Mahogany Road, Route 76

W-E (E-W)

is a native of Southeast Asia, cultivated and naturalized world-wide in the tropics. The leaves are deep green, 1-2 inches long, and the branches have thorns. When ripe, the edible fruits are about 1/2-inch in diameter, have maroon red skin with faint dots, and a sweet citrus flavor. Beware of the green seeds which are bitter and large. One can hear **Bridled Quail-Doves** at many times of the day. The gut has pools with mollies and towering trees draped with vine roots. There are **Yellow-crowned Night Herons and Green-backed Herons** when the pools have water. The **Golden Silk Orb Weaver Spiders**, which make glistening golden webs, are common. The waterfall described at mile 4.3(5.35) is upstream about 0.3 mile. The most serene time to hike is on the week-end, when the Springfield Rock Crusher at the quarry ahead is not operating.

<0.1>

4.7 (4.95) End of **"rain forest" drive**

<0.35>

5.05 (4.6) Rock quarry on left or N produces crushed rock from the Judith Fancy and Cane Valley Formations.

<0.25>

5.3 (4.35) Paved road to right or S "To 705" near the green mileage sign "5". This road goes S 1.1 miles to Centerline, Route 70. [The road is described on p. 35, mile 10.4(5.0).] Along the way are **ruins of Estates Grove Place and Plessen**, a triple-trunked Baobab Tree, and a mural.

<0.15>

5.45 (4.2) Junction with paved road to N and S at a stop sign. Go straight or E to continue on the Mahogany-Midland Side Trip. Route 76 turns right or S here.

Remainder of Route 76 (N-S):

0.0 Junction with Mahogany Road at mile 5.45(4.2). On the left or E across from a bar and restaurant sign less than 0.05 mile south of this junction is a **Breadfruit Tree**. This distinctive tree has large, dark green leaves about 1.5 feet long in a feather-like form. It is a seedless variety, which must be propagated from root cuttings or layering. It was introduced from its native South Pacific to the West Indies by the British (under the famous Captain Bligh of *H.M.S. Bounty*) and the French in the 1700s to supply cheap food for the slaves.

0.2 Paved intersection. Continue straight or S.

0.45 Junction with Route 705 and end of road. Turn left or E to continue to Centerline. Driving straight or S onto dirt at this intersection, one goes along the W side of a school fence. At the SW corner (0.15 mile) the road continues into a field which has **ruins and mills of Estate Plessen**, about 0.1 mile further south.

Side Trip Along Mahogany Road, Route 76, and Midland Road, Route 72

Mahogany Road, Route 76 (concluded)
River Road, Route 69
Midland Road, Route 72

Mahogany Road, Route 76
W-E (E-W)

Remainder of Route 76, N-S (continued)

0.8 Junction with River Road, Route 69 at mile 3.8(0.6). Turn right or S here to go 0.6 mile to Centerline, Route 70 [p. 33, mile 9.8(5.6)].

<0.5>

River Road, Route 69

5.95 (3.7) Paved road to right or S at a stop sign. At this point you will begin a short distance on River Road, Route 69. Go straight or E to continue. [See "Side Trip on River Road", p. 135, mile 3.5(0.9)]. Centerline is 0.9 mile to right or S [p. 33, mile 9.8(5.6)].

<0.2>

6.15 (3.5) Veer left or N

<0.05>

6.2 (3.45) Square-shaped **chimney of Estate Lower Love** is in the distance to right or S of road.

<0.05>

Midland Road, Route 72

6.25 (3.4) Junction with Midland Road, Route 72 on right or E. To continue make a right turn to E. This is mile 3.15(1.25) for Route 69 [p. 135], which continues straight or N here. The two largest trees on the left or N fenceline just E of this junction between the power pole and the sign "72E" are native **White-cedar Trees**. They are characterized by multiple trunks with corded, grayish bark. The flowers are papery, pink trumpets, which cover the tree during the main flowering in spring.

<0.5>

6.75 (2.9) Driveway to Annaly Farms. The large trees on the left or north side of the road here can be identified in sequence and are very nice specimens. From W to E: **Mammee Apple, Mango Tree** at green mileage sign ".5", a small **Hogplum, (chimney of Estate Upper Love)**, **Strangler Fig** with buttress trunk, and an immense **Tibet**.

<0.05>

6.8 (2.85) -Dirt road to left or N. Then in order from W-E: a small **Raintree or Samán**, four **Puerto Rican Royal Palms**, and a **Flamboyant Tree**, native to Madagascar and introduced world-wide in the tropics for its marvelous orange-red flowers.

<0.05>

6.85 (2.8) Holy Cross Episcopal Church on the left or N side of a junction. Veer left or N around the church. The paved road to right or S "To 70" goes 0.95 mile to Centerline.

Side Trip Along Mahogany Road, Route 76, and Midland Road, Route 72

Midland Road, Route 72

W-E (E-W)

"To 70":

N-S (S-N)

0.0 (0.95) Junction with Route 72, mile 6.85(2.8)

0.6 (0.35) Dirt road to left or E

0.9 (0.05) Dirt road to left or E goes past a **freshwater pond with birdwatching opportunities** on the N at (0.05 mile) and **ruins of Estate Castle Burk** at 0.1 mile. It ends at mile 1.4 of an access road to **ruins of the sugar factory at Bethlehem Old Works** [described with a map in "Centerline Tour", p. 31, mile 7.75(7.65)].

0.95 (0.0) Junction with Centerline [p. 33, mile 9.0(6.4)].

<0.1>

6.95 (2.7) **Water windmill of Estate Jealousy** on the left or W

<0.3>

7.25 (2.4) Road veers right or E

<0.1>

7.35 (2.3) **Mill and ruins of Estate Jealousy** are on a low rise to right or S. A paved road to left or N is a 0.65 mile loop into a residential area, built among **ruins of Estate Coble**.

<0.15>

7.5 (2.15) The tall, round **chimney of the sugar factory at Bethlehem Old Works** is on the right or S side of the road.

<0.15>

7.65 (2.0) **Cruzan Gardens Nursery** is on the left or N: telephone 778-0700, hours M-Sa, 8am-5pm. This nursery specializes in **native plants of St. Croix** and other tropicals. An old **watermill of Estate Colquehoun** is on the nursery grounds, which are located along Bethlehem Gut.

<0.15>

7.8 (1.85) Paved road to left or N into Estate Colquehoun residential area. At 0.3 mile there is a junction with a dirt road. **Ruins of Estate Colquehoun** are on the NW corner. The round rubble pile is a former grinding windmill and the belltower still has some of its mountings. The plantation bell was used to signal important times of the day on the old estates.

-At 0.5 mile pavement ends. One can hike the road, then up the gut through shady forest and finally open brush to the **ruins of Estate Hermitage**. The cane-grinding windmill and factory are about 0.4 mile away just E of the gut.

<0.1>

7.9 (1.75) Dirt Road to left or N into Estate Mt. Pleasant residential area. A **Breadfruit Tree** is on the left or N, behind a red brick wall of a corner house. [See p. 85, mile 0.0 for information on this species.]

<0.4>

8.3 (1.35) Paved road to left or N into Estate Mt. Pleasant residential area. At 0.3-0.45 mile there is a large orchard of **Mango**

Midland Road, Route 72
W-E (E-W)

Trees on the right or E side. Blue Mountain (1096 feet) with its radio towers and antenna is just north of this area.

<0.45>
8.75 (0.9) Junction with Canaan Ridge Road, Route 73 [p. 132, mile 3.3(1.4)]. A left turn to N goes 3.25 miles to the village of LaVallee on the north shore. A right turn to S goes 1.4 miles to Centerline.

<0.2>
8.95 (0.7) Border of **Pencil Trees** on left or N of road for next 0.05 mile. They are native to Africa.

<0.2>
9.15 (0.5) **Friedensfeld Moravian Church** on the left or N. It was partially prefabricated in Germany in the mid-1800s. The old wooden building NW of the church is the original manse.

<0.4>
9.55 (0.1) Paved road to left or N next to St. Croix Cock Pit; cock fights are regularly held here. At 0.25 mile one comes to the end at a housing complex built near the **arch and mills of Estate Glynn**. The road to right or E at the arch gives a close approach to the mills.

<0.1>
9.65 (0.0) Junction with North Side Road, Route 75 [p. 99, mile 4.75(1.55)]. A left or N turn goes 4.75 miles to the Golden Rock Shopping Center outside Christiansted. A right or S turn takes you 1.55 miles to the La Reine Shopping Center on Centerline.

SIDE TRIP ALONG CREQUE DAM ROAD, ROUTE 58

(Also Sprat Hall Valley Road, Annaly Road)

"Creque" is pronounced "creaky." This unpaved road is on the side of a watercourse, flanked by some of the largest trees on the island, some over 100 feet tall. The habitat is sometimes called "rain forest," but it is really secondary growth forest (not a virgin forest), and it does not receive enough rain to be a true rainforest. Sometimes sections of road are very rough, especially after heavy rains. It was completely washed out by the close pass of Hurricanes David and Frederick in 1979. The road meets the Western Scenic Drive in 3.2 miles. One can circle through the "rain forest" portion by driving up Route 58 to its junction with Route 765 at mile 2.95(0.25), turning right or S to Mahogany Road, then turning right or W and driving through the "rain forest" portion of Mahogany Road, rejoining Route 63 at mile 2.4(4.55).

W-E (E-W)
0.0 (3.2) Junction with Route 63 [p. 72, mile 3.7(3.25)].

Side Trip Along Creque Dam Road, Route 58

W-E (E-W)

<0.1>

0.1 (3.1) Paved road north into Sprat Hall Plantation. [See Route 63, p. 73, mile 3.75(3.2) for a description of this stop.]

<0.4>

0.5 (2.7) Concrete pad for stream crossing, guard rail on left or W. A **Hogplum Tree**, with a trunk over 2 feet in diameter, is on the SE corner of the pad, and as you cross over, a **Tamarind Tree** is on the NE corner. The Hogplum is native, the Tamarind from the Old World tropics.

<0.55>

1.05 (2.15) On the left or N, about 50 feet W of the green mileage sign "1", is a magnificent **Kapok Tree**. Most of the trees with buttressed trunks you will see along this road are Kapoks. Note the heavy thorns on the trunk. Another common tree on this road with thorns on its trunk is the **Sandbox Tree** (also Monkey-pistol). The Kapoks have thorns that seem irregularly spaced along the trunks compared to the dense, closely-packed arrangement on Sandbox Trees. The easiest way to tell the two trees apart is to look at the leaves. Kapoks have their leaves arranged like a fan, with about 5 leaflets radiating from a central stalk. Sandbox have simple leaves about 2 inches long and nearly as broad, tapered on the end. A good example of a Sandbox tree is at Creque Dam, mile 1.3(1.9) ahead.

-Across the road and streambed from the Kapok is a **grove of Cacao Trees**. The Cacao Trees have spindly trunks and are less than 20 feet tall. To prosper they need the shade of taller trees as you see them here. The tapered, oblong leaves are about 1 foot long and up to 4.5 inches wide, usually broader than mango leaves. New growth at the branch tips is reddish-bronze. The fruit is shaped like an acorn squash and usually grows right out of the main trunk. It is golden-colored when ripe. Chocolate is made from the beans inside.

<0.05>

1.1 (2.1) On the left or N at about 10 yards uphill of a small bridge over Creque Gut is a **Mammee Apple Tree** with a trunk about two feet in diameter. It has a characteristic erect growth pattern and an orange tinge to the bark which flakes in roundish pieces. The leaves are glossy, dark green, about 5 inches long. The fruit is round, 3-10 inches in diameter, with tan skin. The edible part inside is orange pulp surrounding a large seed.

-On the N about 30 yards uphill from the bridge are three large **Kapok Trees** with buttressed trunks.

<0.2>

1.3 (1.9) Creque Dam, a concrete arch dam built in 1926, is 45 feet high. It was built to supply potable water to Frederiksted under the auspices of the U.S. Navy when it was in charge of the V.I. It was designed to hold 9 million gallons from

W-E (E-W)

a 400-acre drainage. At the present rate of siltation, it may be filled before it is 100 years old. Looking on the two sides of the dam when it has no water will give you an idea of how much the dam has filled in since 1926. In 1928 a hurricane hit the west end of St. Croix with full force, causing tremendous damage. This dam was topped by a wall of water 8 feet high during that hurricane.

Fig. 32. Curtains of Spanish Moss grace the trees at Creque Dam.

-This is a good place to appreciate the height of the trees in the moist forest. A tall **Kapok Tree** with its five-part leaves is the first (most southern) tree on the downstream side that projects 40 feet above the dam top. Two large native **Sandbox Trees**, their trunks covered with spines, also project 30-50 feet above the dam top. The flowers, which are most abundant in winter through summer, are easy to see from this vantage. The maroon male flowers are pendulant, elongate, grape-like clusters; the female flowers are fleshy, maroon, and trumpet-shaped, about 1 inch across. The Sandbox Tree gets its name from its fruits, which were hollowed out when green and used to hold sand that was sprinkled on letters to blot ink. The name Monkey-pistol Tree also derives from the fruits, which release their seeds by exploding. During the summer when the fruits dry, one can hear the explosions followed by the patter of the seeds falling through the leaves to the forest floor. Many of the trees surrounding

W-E (E-W)

the dam have **Spanish Moss**, a bromeliad epiphyte. The trees in the Creque Gut have many kinds of **epiphytes**, plants which use the trees as support rather than growing in the ground. The principal groups of epiphytes here are ferns, bromeliads, and fungi, but only the fungi are parasites. On some of the old trees little bark is showing, the epiphytes are so thick.

-To really appreciate the forest, you should get out of your car and **hike** through it to the concrete pad at mile 1.75(1.45). Please walk, look, touch, smell, and listen. The round trip for the walk will be about 0.9 mile. The rich, loud whistles you hear in the forest are from the **Pearly-eyed Thrashers**, which sing all day long in areas as shady as Creque Dam.

<0.35>

1.65 (1.55) Concrete pads, two about 30 yards apart.

<0.1>

1.75 (1.45) Concrete pad where road crosses the stream. (For those going E-W, this is a good place to start a walk through this lovely forest. The round trip to Creque Dam is 0.9 mile.)

-Three large trees are just W (downhill) of the pad. At 100 feet is a **Kapok Tree** [discussion at mile 1.05(2.15)], at 80 feet W is another **Kapok**, and at 50 feet W is a large **Mango**. The Mango has distinctive long, feather-shaped leaves. Large mangos tend to develop very knobby trunks, but have smooth bark with few grooves. Some people may say "Wait a minute. That tree at 80 feet W is also a mango!" What has happened here is that a smaller mango has grown very close, or perhaps intertwined its roots with the Kapok. In a tropical forest, one must often follow any particular trunk all the way to the branches with leaves, and observe the leaves to determine a trunk's identity. In this case the main trunk leads to the palmate leaves of a Kapok Tree. A common occurence in tropical forests is that trees of different types have interconnected root systems, as may have happened here. This strategy probably enables trees to grow in denser concentration.

<0.2>

1.95 (1.25) **Ruins of greathouse of Estate Mount Victory** on left or N. This was converted to a school in the 1840s. An old carriage bridge still remains across the road from the ruins. Another crumbled structure across from the greathouse is remains of a watermill.

<0.05>

2.0 (1.2) Dirt roads to left and right. The right or S road over an old carriage bridge is the original road that connected these northern estates to Frederiksted. (There was no Creque Dam Road.) Now it goes near **ruins of Estate Punch** but is no longer through to Mahogany Road.

-The road to left or N goes to Estates Mount Victory and Nicholas.

Side Trip Along Creque Dam Road, Route 58

W-E (E-W)

Nicholas Road (S-N):

0.0 Junction at mile 2.0(1.2) with Creque Dam Road

0.1 Fork. Turn left or W to continue to Estate Nicholas, but you may need permission. A right turn goes to **Mount Victory ruins and mill.** To visit Mount Victory ask at the house for permission to walk to the ruins, which are along the road a short distance past the house. The factory ruins and chimney are on the right or E just after you turn N and begin climbing uphill. One can follow the road uphill to the mill (which has been walled closed for use as a cistern) for a lookout on the peaceful surroundings. Portions of the covered aqueduct for the cane juice leading south to the chimney and factory are still visible.

0.6 Impressive **factory ruins of Estate Nicholas** (private) on right or E as road swings left or W. The mill is on the northern hill above the factory.

<0.1>

2.1 (1.1) On the right or S are some large trees with trunks about 3 feet in diameter. Going uphill they are **Kapok, Raintree, Raintree, Sandbox Tree**. The green mileage sign "2" is at the base of one of the Raintrees. [See p. 89, mile 1.05(2.15) for discussion of the Kapok and Sandbox Trees.] The Raintree is native from Mexico to Brazil. It is introduced and naturalized on St. Croix. It tends to have the huge and spreading habit that you see here.

<0.85>

2.95 (0.25) Junction with Route 765, paved road to right or S [p. 84, mile 1.2]. This road goes 1.2 miles S to join Mahogany Road (Route 76).

<0.25>

3.2 (0.0) End of Creque Dam Road at the junction with Western Scenic Drive, Route 78 [p. 122, mile 16.15(3.35)]. A left or W turn here takes one 3.35 miles to Ham Bay on the West End Road (Route 63). The western section of road can be severely eroded and passable by four-wheel drive only. A right or E turn goes along the highest road on the island, with many scenic views, 4.45 miles to River Road.

NORTH SHORE TOUR
INCLUDING PARTS OF ROUTES 75,80,69, AND 78 WEST

This tour is one of the most scenic drives on St. Croix. Miss it at your own risk! It follows the northwestern coastline as closely as major roads will allow. (On the "Map of Maps", it is the tour outlined in green along the northern shore.) This guide indicates most of the access points to the beaches and diving along the northern coast, and access to one of the largest mangroves on St. Croix at Salt River. The final western part of the tour goes along the top of spectacular Maroon Ridge, which drops fairly abruptly about 700-800 feet to the sea. The tour starts from Christiansted Golden Rock Shopping Center, but one can take it from Frederiksted by driving along the western shore on Route 63 to the entrance of Ham Bay and starting the tour backwards (W-E) there. The tour uses all of the main roads along the northwest shore of St. Croix. It travels on North Road (Route 75) for 3.7 miles, then turns onto North Shore Road (Route 80) and follows the entire length of Route 80 for 7 miles along the northwestern coastline. Route 80 terminates in Route 69, and this tour uses the next 0.75 miles of Route 69 to the junction with Scenic Drive West (Route 78). Turning onto the Scenic Road, this tour then travels Route 78 for 7.8 miles along the Maroon Ridge spine to paved road near Hams Bluff of the West End Road Tour (Route 63).

Because of the complex system of several Routes for this tour, two sets of mileages are given. The larger print gives the mileage of the tour from beginning to end as usual. The smaller print just below gives mileages for each individual Route by itself.

North Road, Route 75 (also North Side Road)
E-W (W-E)

0.0 (19.5) Beginning of Route 75 at traffic light outside of
0.0 (6.15) Christiansted at junction with Centerline Tour, Route 83 [p. 21, mile 1.4(14.0)]. **Golden Rock Shopping Center with a supermarket** is on the SW corner of this junction.

<0.3>

0.3 (19.2) Road to right or N. There is a dark gray stone façade on
0.3 (5.85) the building on the NE corner at this junction. One can use this road to see the **mill of Estate Golden Rock**. The **mill** is surrounded by Sugar Beach Condominium. Go N 0.25 mile to the road end at a stop sign and turn left. The Condominium is 0.15 mile further, on the seaward side.

<0.05>

0.35 (19.15) Road to left or SW between McDonald's and a car dealer.
0.35 (5.8) This road leads to a nice **view of Christiansted** and the **interior secondary growth forest**. You may continue on a rough section all the way to the **Experimental Forest of Teak and Mahogany Trees**, but four-wheel drive is required for this last part.

Back Road to Experimental Forest:
E-W (W-E)

0.0 (1.7) Junction with Route 75 at mile 0.35(19.15). A road sign at the start of the steep uphill section says "Little Princesse Hill Road."

NORTH SHORE TOUR
INCLUDING PARTS OF ROUTES 75,80,69, AND 78 WEST
North Road, Route 75

Back Road to Experimental Forest (continued)

E-W	(W-E)	
0.15	(1.55)	Junction with a **view of Christiansted Harbor and Buck Island** between trees. Continue straight or SW.
0.3	(1.4)	Fork. Veer left or SE following the main road.
0.45	(1.25)	Turn left or W uphill at sign "11C". Road changes to concrete here.

Fig. 33. Collecting coconuts on Little Princess Beach.

0.55	(1.15)	End of pavement as you turn left or SW onto dirt. The uphill sign here is "Bellevue". Now

Back Road to Experimental Forest (continued)
E-W (W-E)

		the road passes through **secondary growth forest** that replaced the sugar cane plantings. It is a delightful, shady drive.
0.85	(0.85)	House on left or SE side of road.
1.0	(0.7)	Junction. Turn right or NW.
1.1	(0.6)	Dirt road to N for four-wheel drive goes 0.2 mile to end at clearing overlooking peninsula of Judith's Fancy and St."C" Condominium.
1.15	(0.55)	Fork. Go left or S.
1.35	(0.35)	Four-wheel drive is required beyond a dirt road to right or N.
1.7	(0.0)	Cattle guard and entry into Experimental Forest. [To tour through the forest and exit on Centerline at mile 4.35(11.05): see "Side Trip to Experimental Forest", p. 45, mile 3.5(0.0) and proceed following mileage for (W-E) from this point.]

<0.35>
North Road, Route 75 (continued)
E-W (W-E)

0.7 (18.8) Traffic light junction of Route 75, Route 74, and road
0.7 (5.45) north to Little Princess **beach area**. Veer left on Route 75, the main or middle road to the NW.

Little Princess Beach area:
Turn right or N at traffic light. (You will actually be going nearly straight, as Route 75 veers slightly left or NW at the traffic light. Extreme right goes to housing.)
S-N:

- 0.0 Junction with Route 75 at mile 0.7(18.8)
- 0.25 Continue straight or N onto dirt road as pavement veers to left or W.
- 0.65 Turn right or E along **white sand beach with many coconut palms**. Beach access here is to western end of Long Reef which forms Christiansted Harbor.
- 0.85 Road ends.

Turn to extreme left or SW at the traffic light to travel Route 74.

Route 74 (E-W):

- 0.0 Junction with Route 75 at traffic light, mile 0.7(18.8)
- 1.3 Paved road to left or S is Route 81, running 1.9 miles south to join Centerline.

Route 81 (N-S):

- 0.0 Junction with Route 74 at mile 1.3
- 0.2 Veer to right or SW.

Route 81, N-S (continued):

0.55 Route 81 veers S. A paved road runs uphill.

0.9 View of **ruins of chimney, greathouse, and mills of Estate Sion Hill** on hill to left or E

1.7 Veer left or E

1.8 Turn right or S. On the left or NW corner of this intersection is a school built around the **mill and ruins of Estate Sion Farm**. The school has incorporated the ruins into its structure simply by replacing the mortar and adding on. These ruins are fine ones to walk among. The library is in the old boiling station. Remains of an old bell tower are on a building now connected to the remains of the sugar grinding mill. The many **ferns on the old block walls** are particular to limestone.
-A left or N turn here would take you into the Experimental Forest. [See "Side Trip to the Experimental Forest," p. 44, mile 2.9(0.25).]

1.95 Junction with Centerline (Route 70) [p. 26, mile 4.35(11.05)]. Jog to E, then S again to continue Route 81.

2.55 Junction with Route 68 [p. 53, mile 2.45(0.4)].

Continuing Route 74 (E-W):

1.4 Turn right or N downhill to view a nicely cleared **old watermill of Estate Belvedere** in a private front yard at 0.15 mile on the left or W. You can continue on this sideroad 0.55 mile to the saddle and view below.

1.9 Saddle between two peaks with a **view of both north and south shores at middle of island.**

2.4 **Ruins, mill, and chimney of Estate Rattan**, incorporated into a private home on right or N

2.5 End of Route 74 at Route 79. [See p. 98, mile 0.75(1.6) of Route 79.]

<0.45>

Continuing on Route 75 from traffic light at 0.7(18.8):

E-W (W-E)

1.15 (18.35) Theodora Dunbavin School on the right or N. This is the
1.15 (5.0) only **19th-century Danish schoolhouse** still used as a
<0.25> school.

1.4 (18.1) A right or N turn at the Texaco station leads to **beach**
1.4 (4.75) **access**. (S-N):

0.0 Junction with Route 75 at mile 1.4(18.1)

0.15 **Ruins and mill of Estate LaGrande Princess**

0.35 Junction at a stop sign. Behind you to the S on a hill above the ruins is the **mill of LaGrande Princesse**. Continue straight or N 0.1 mile onto dirt for public access to **Pelican Cove's white sand beach,**

North Road, Route 75
E-W (W-E)

with coconut palms for shade. The Cormorant Beach Club, which fronts most of the beach, has its entrance at 0.15 mile on the road to right or E. Public access is a 7-foot-wide strip at the far western end of the hotel grounds (no signs). This area is a **good snorkeling spot.** It lies at the extreme western end of Long Reef and provides a large shallow area (chest height for adults) among coral heads and rock ridges.

<0.15>
1.55 (17.95) **Teak Tree** is on the SW corner of a fenced playground on
1.55 (4.6) the right or N side of the road. There are extensive groves of Teak planted during the 1950s and 1960s on the western half of St. Croix. However, teak does not grow well here, probably because rainfall is insufficient.

<0.35>
1.9 (17.6) Paved road to right or N, Route 751, goes to Estate Judith's
1.9 (4.25) Fancy (long ago known as the Hemer Peninsula) and to Salt River Bay, which has extensive **mangroves, seagrass beds, beach access, and a surfing area.**

Route 751 to eastern side of Salt River Bay (S-N):

0.0 Junction with Route 75 at mile 1.9(17.6)

0.35 A left or W turn here onto "To 75" goes by the **chimney, mill, and factory ruins of Estate St. John** at 0.25 mile. This sugar factory was a cooperative used after emancipation of the slaves. It never did very well, chiefly because it was no longer economical to grow sugar on St. Croix and compete in the world market with sugar beets.

0.55 Sign "End of Route 751". Begin Hamilton Way at guardhouse entrance into Estate Judith's Fancy.

0.65 Junction with Jefferson Way to the right or E. This road runs 1.3 miles along the eastern shoreline of the peninsula, with several turnouts for **beach access.** It joins Grapetree Road at 1.0 mile, which connects to mile 1.7 below.

1.1 **Ruins and mill of Estate Judith's Fancy** incorporated into a private residence on the right or E. The ruins are speculated to include remains of the Governor's Chateau of 1659 and are sometimes called the "French Ruins" or "Maltese Ruins".

1.6 Hamilton Drive ends and the road veers right or N onto Sea Grape Way. A left or S turn onto Robin Lane provides a **good aerial view of Triton Bay** at 0.45 mile. Due to chain barriers one must walk this road. Triton Bay is the easternmost arm of the Salt River estuary, a mangrove owned by The Nature Conservancy.

Route 751 to Salt River Bay, S-N (continued)

1.7 Pavement ends. The beach area of Salt River beyond this point is being developed, and access will change. Continue on foot 0.3 mile to left or NW to beach and estuary. A right or E turn here onto pavement of Grapetree Road goes 0.2 mile to Jefferson Way, mentioned above at mile 0.65.

In Salt River Bay:

-The best **surfing** is in the mouth of the Salt River Bay. Snorkeling is mostly over seagrass beds.

-Just offshore is the fast-rusting **wreck of the interisland freighter** *Cumulus*. The *Cumulus* was bound "down island" with a cargo of stolen cars in October, 1977, when it went aground here in stormy seas on Saltriver Reef.

-The northern tip of the peninsula is Saltriver Point, and the shoal area with breaking waves 0.25 mile offshore is called Whitehorse.

<0.8>

Continuing on Route 75:

E-W (W-E)

2.7 (16.8)
2.7 (3.45) Paved road to right or N "To 751" leads past **ruins of sugar factory at Estate St. John** [p. 97, mile 0.35].

<0.05>

2.75 (16.75)
2.75 (3.4) Junction with Route 79.

-The road to N goes 1.15 miles through a residential area to Triton Bay, an arm of the Salt River Estuary. Triton Bay is a **preserve of The Nature Conservancy**. The road ends at the Dyck Marine Center, administered by Fairleigh Dickinson University. There are no visitor facilities. This station is preparing an underwater research and living habitat. Teams of diving scientists will spend a week living in a pressurized chamber about 50 feet below the surface. Living at the pressure of the surrounding water enables the divers to work long periods of time at depth. By remaining at high pressure and not coming to the surface, the divers avoid having to take frequent, long surface intervals to prevent decompression sickness ("the bends"). Decompression procedures are done only once, at the end of the week-long diving period.

-The road to S goes past the **Island Center for the Performing Arts**, an outdoor auditorium, and the **St. Croix Hospital**, ending at Centerline in 2.35 miles.

Route 79, Pepper Tree Road:

N-S (S-N)

0.0 (2.35) Junction with Route 75 at mile 2.75(16.75)

0.75 (1.6) Junction with western end of Route 74. [See p. 96, mile 2.5 on Route 74 for the log of this road.] The **chimney of Estate Rattan** can be seen on a hill to SE of the junction.

INCLUDING PARTS OF ROUTES 75,80,69, AND 78 WEST

North Road, Route 75

North Road, Route 75
E-W (W-E)

Route 79, Pepper Tree Road (continued):
N-S (S-N)

1.9 (0.45) **Island Center for the Performing Arts** on the right or W

1.95 (0.4) **St. Croix Hospital** on left or E

2.35 (0.0) Junction with Route 70 [p. 27, mile 5.0(10.4)].

<0.2>

2.95 (16.55)
2.95 (3.2) Junction with dirt Dolby Hill Road to residential area on right or N. The private house on the NE corner is built into the **ruins and mill of Estate Dolby Hill** (also called Montpellier).

<0.85>

3.8 (15.7)
3.8 (2.35) Junction with North Shore Road (Route 80). Turn right or N onto Route 80 to continue this tour, and skip to p. 101, mile 3.8(15.7). Several large **Flamboyant Trees** covered with bright orange-red flowers in spring and summer are on the left or W side of the road.

-North Road (Route 75) continues 2.35 miles SW to join Centerline.

Remainder of North Road, Route 75
E-W (W-E)

3.8 (2.35) Junction with Route 80 at mile 3.8(15.7)

<0.7>

4.5 (1.65) **Watermill, cane-grinding mill, and ruins of Estate Glynn** near housing on right or W

<0.25>

4.75 (1.4) Junction with eastern end of Route 72 (Midland Road) which runs into Mahogany Road (Route 76) on its western end. [See p. 88, mile 9.65(0.0).] The **mill of Estate Bonne Esperance** is visible on a hill to the left or E between this junction and the next point.

<0.35>

5.1 (1.05) Nursery stock plantings on the left or E just south of the car dealership fence. This is a good place to see some uncommon native plants as well as exotics that are used in horticultural plantings.

Horticultural plantings S of Cruzan Motors fence. (Row numbers start from fence on N [fenceline = 1].)

Row 1. The trees with glossy green leaves and red figs (in summer) are the **Benjamin Fig**, a native of SE Asia. The trees with silvery green leaflets that are grouped into palm-like clusters of five have showy clusters of yellow flowers in the late summer and fall. They are the **Silver Tabebuia Tree**, a native of Paraguay and Argentina.

Row 2. This row is a good place to distinguish several kinds of palms. First come the **Royal Palms** with the wavy leaflets distributed along

Cruzan Motors horticultural plantings (continued)
a long blade. The clusters of leaves look somewhat like a fountain spray. **Caribbee Royal Palms** predominate in this row and in other rows further south. These palms usually have a trunk that is enlarged at the base and then gradually reduced in girth, in contrast to St. Croix's native Puerto Rican Royal Palm. (The Puerto Rican Royal has a swelling midway along the trunk looking like a snake that has just swallowed a bolus.) The **Manila Palms**, native to the Philippines, are very short in comparison to the Caribee Royals and have red dates in summer. A small palm with a thin, delicate trunk (thinnest of all in these rows) is the **Alexander Palm**. Also scattered about in this row and some of the others are three species of **Fan Palms**, with their leaves looking like the kind of air-conditioning used by Cleopatra. The leaflets tend to radiate from the blade nearly at the same point, giving them the fan shape fronds. From California comes the **Desert Fan Palm** with the fronds well separated and stiffly flat. The attenuated leaflet ends hang very gracefully in an orderly way. A similar palm, but with the fronds undulating, may be a native to St. Croix. This is the **Puerto Rican Palmetto**, which is most abundant locally in the Christiansted area. On Puerto Rico it is used for straw hats, baskets, mats, and hammocks. The last fan palm is the most striking. Its leaves have diameters of 2-3 feet and do not have the trailing tips of the two others. These are the **Fiji Fan Palms**, which you will not find (as long-lived specimens) outside of the tropics. It is native to Pacific Islands. You will also see some **Date Palms** scattered among these plantings. Date palms have very long leaves or fronds that have rather stiff leaflets. (The leaflets do not wave in the breeze.) The plants with many small red spines along the borders of strap-like leaves are the **Common Screwpine**, a native of Madagascar. The swirling arrangement of the leaves gave this tree its name. These are relatively young, and although they look like agaves now, they could someday grow to 60 feet and have a trunk!

Row 3. This row has the palms of #2, as well as two other non-palm trees. Near the road are **Black Calabash**, an uncommon native tree of the moist forests, with dark green, broad, elliptical leaves, 3-6 inches long. The summer fruits are slightly elliptic, fist-sized gourds. The seeds are edible, and used in a beverage. When ripe, the little gourds are surprisingly light, and are spread by the ocean currents. Hence this tree is widespread in tropical America. Near the center of this row is a tree with small (1-inch diameter) heart-shaped, grey-green leaves. This is a veneer-wood tree from India, the **Indian Rosewood**, attractive, salt and drought resistant, and extremely tolerant of a wide variety of growing conditions. Unfortunately, it is highly susceptible to wood-borers, who go wild when they taste its honey-like sap (which gave this tree its name). Once these borers find a tree, it oozes sap for months or years and then eventually dies.

Rows 4-7 have trees listed above.

Row 8. Near the road are **Sausage Trees** with their large pendulous flowers and sausage fruits. The back has **Indian Almonds** [see p. 33, mile 9.8(5.6)], with their attractive, deep-green leaves. Mixed in with the Sausage Trees towards the front are a few **Sea Hibiscus** (also called Beach Mahoe) trees, with heart-shaped papery leaves, 4-7 inches across and long.

Cruzan Motors horticultural plantings (continued):
They have spectacular yellow hibiscus flowers that turn red with age, and attractive star-shaped seed pods in summer. This tree is very rare on St. Croix. It is found on many tropical seashores, but is thought to be originally from the Old World.

Row 9. Most of the **Fiji Fan Palms** are in this row.

Row 10. Towards the front are **Flamboyant Trees**, which have spectacular red-orange flowers and 1-foot pendulous seed pods, and a native tree that is commonly used in horticultural plantings, the **Pitch-apple**. The apple-like fruits open to reveal yellow seeds in bright red pulp. The copious sap hardens nicely and has been used for boat calking, plaster, and medicine. The triangular, leathery green leaves are so stiff they were used as playing cards by the conquistadores. Towards the front are a few **Sea Hibiscus**. In the final part of this row are some trees with very large leaves 12-18 inches long that attach without a stalk to the branches. This is **Barringtonia**, native to tropical shores from the South Pacific to India. Where native, the trunks are used for dugout canoes, and the waterproof, corky fruits are made into fishnet floats. Crushed bark, leaves, seed, and fruit are used as fish poison.

Row 11. This last row has a good stand of an uncommon tree of moist forests called **Santa-maría**. The deep-green, leathery leaves have a distinctive vein pattern, very close-packed and unbranched like a bird feather. The fruits look initially like a small 1-inch fig, but eventually the exterior dries and becomes hard. It is widely distributed in the West Indies to South America.

<0.2>

Remainder of Route 75 (continued):

E-W	(W-E)	
5.3	(0.85)	Junction with Route 709 on left or E. [p. 28, mile 1.7].
<0.25>		
5.55	(0.6)	Junction with Route 707 to right or W [p. 31, mile 1.45]. This road leads 0.8 mile to **freshwater Fredensborg Pond** and **ruins of Bethlehem Old Works sugar factory.**
<0.25>		
5.8	(0.35)	Kingshill **Post Office** on left or E
<0.05>		
5.85-6.15	(0.0-0.3)	Villa LaReine **Shopping Center with a supermarket and police station** on left or E. "Reine" is pronounced "rain".
<0.3>		
6.15	(0.0)	Junction with Centerline (Route 70) [p. 28, mile 7.0(8.4)].

North Shore Road, Route 80 (also Oxholm Road)

E-W	(W-E)	
3.8 0.0	(15.7) (7.15)	Junction with Route 75 at mile 3.8(15.7),3.8(2.35). If driving this tour (W-E), skip to p. 99, mile 3.8(15.7).

NORTH SHORE TOUR
INCLUDING PARTS OF ROUTES 75,80,69, AND 78 WEST
North Shore Road, Route 80

E-W (W-E)
<0.05>

3.85 (15.65) Dirt roads to left and right just before (S of) crossing a
0.05 (7.1) bridge over Salt River. Right or E goes 0.1 mile to a **coconut grove** with very tall trees.

-Left or W goes to Glynn Farms area and 1.65 miles to Route 73.

Glynn Farms Road

E-W	(W-E)	
0.0	(1.65)	Junction with Route 80 at 3.85(15.65)
0.05	(1.6)	Large **West Indian Mahogany Trees** line the S side of road.
0.15	(1.55)	Road forks. Go right or N. On the left or S fork is a residential area amid **ruins of Estate Concordia. Mill, chimney, and other buildings** in good condition.
0.5	(1.15)	Dirt road to right or N to **ruins and mill of Estate Windsor** (private)
0.7	(0.95)	**Region of farms of Estate Glynn**. Dirt road to left or S goes past St. Croix Produce (open M-F, 12-6 PM; Sa 8-12 AM) and fields of **bananas, coconuts, soursops**, and **sugar cane**.
1.0	(0.65)	Junction. Dirt road to right or N has remnants of **old cobble road** leading to Estate Libanon. Road ends at a fence at 0.15 mile.
1.6	(0.05)	Dirt road to right or N goes to water tank for Mon Bijou and past **chimney and ruins of Estate Libanon**. Road ends at 0.2 mile. -Concrete road to left or S 50 yards to W of this old dirt road goes to Mon Bijou. About 15 yards S from the main Glynn Farm road and near this paved road is a line of four large, native **Gregre Trees**. Gregre produces a wood so durable it has been used for railroad ties, bridges, and marine pilings.
1.65	(0.0)	Junction with Canaan Ridge Road (Route 73) [p. 132, mile 2.7(2.0).]

<0.3>

North Shore Road, Route 80 (continued)
E-W (W-E)

4.15 (15.35) Paved road to left or W is start of Eastern Scenic Drive,
0.35 (6.8) Route 78 [p. 127, mile 0.0(15.0)]. On the right or E side of the road is a trash collection point, with the **watermill of Estate Morningstar** just E of it in the field.

<0.1>

4.25 (15.25) View of **watermill, chimney, and cane-grinding windmill of**
0.45 (6.7) **Estate Morningstar** on the right or E on hills in back of you (to S). These are easiest seen going (W-E).

North Shore Road, Route 80
E-W (W-E)
<0.35>

4.6 (14.9)
0.8 (6.35) Access point to **mangroves of Sugar Bay**, an arm of the Salt River Estuary. The access point is south of the green mileage sign "6", where there is parking on the left or W of the road, about 200 feet north of the access point. At the access a telephone pole is on the W side of the road and a **Gregre Tree** with a trunk diameter of almost 2 feet is on the E. This tree is riddled with lines of holes made by a woodpecker which rarely migrates to St. Croix, the **Yellow-bellied Sapsucker**. The birds make the holes to cause the sap to flow and attract insects. They later return to the holes and eat the insects. The Sapsuckers favor trees that are particularly resinous, like the Gregre.

-Walk in east 50 feet. At the **Black Mangrove Trees** (with roots protruding straight up out of the ground), turn right or S and walk with the Black Mangroves on your left and thorny acacias on your right towards the SE, wending through Acacia and Black Mangroves. Stay on firm mud, if possible. This is a good area for seeing **wintering warblers** and **Kallaloo Crabs**. The crabs live in the large mud holes that pock the ground.

Fig. 34. As the mangroves of St. Croix are obliterated, so goes the Swamp Fern. A few are left in the Sugar Bay mangrove. Its large size makes it particularly striking.

North Shore Road, Route 80
E-W (W-E)

-At about 0.2 mile you may notice some concrete pens, ruins of an old fish hatchery, and you will then know you are close to a road on a raised berm. At the southern end of the old fish ponds is an area with very large ferns that could serve as background for dinosaur movies. These are native **Swamp Ferns**, which do very well in brackish water. Krause Lagoon was once carpeted with this fern.

-Walk left or N on the berm road about 0.2 mile past thick Black Mangrove forest. You will deadend at a picturesque area of **dense Red Mangrove**. **Large swimming crabs and mangrove oysters are common** in the water sheltered under the mangrove roots. Return the way you came.

Fig. 35. Red Mangrove forest at the end of Sugar Bay.

<0.8>

5.4 (14.1) Paved road to right or N to **site of Columbus' Landing and**
1.6 (5.55) **white sandy beach**. A sign says "Salt River Marina".

Columbus' Landing (S-N):

0.0 Junction with Route 80 at mile 5.4(14.1). To the right or E at the junction is the road to Salt River Marina. The marina has been made by cutting out large sections of mangrove, but there are still **islets of Red Mangrove** visible from the marina road.

Columbus' Landing (continued)

0.1 View of **mill of Estate Salt River** on a hill to S behind you. This mill is best viewed when you return to the Route 80.

0.3 Fork. Turn right or E.

0.4 Monument for National Historic Landmark on the left.

-At 30 yards further is a dirt road to right. Since it is rutted and only 0.05 mile to parking area at end, it is easiest to park at Columbus' Landing beach and walk back here. One can **hike along the shore at the end of this road through Red Mangrove Trees** both to N and S. The N path leads along shore to beach and area described below. The area of mangroves near the parking has many **crab holes**, often with box traps over the openings. This **Kallaloo Crab** is common in mangroves, and is the preferred crab for a delicious soup called "Kallaloo".

Fig. 36. Native trap for Kallaloo Crab, the tastiest resident of the mangrove.

0.45 Turnaround loop and parking for Columbus' Landing. Some of Columbus' men reputedly landed at this point on his second voyage on 14 November 1493. They did land somewhere on the island, and Salt River fits descriptions in the log. He named the island "Sancta Cruz" (Holy Cross), and the name still used by Spanish speakers is "Santa Cruz". Somehow

Columbus' Landing (continued)

the French equivalent "Saint Croix" stuck as the official name. The Caribs, of course, had a name for the place, "Ay-ay", which Columbus recorded but dutifully disregarded.

-This site is an archaeological treasure on St. Croix, although it is not treated as such. Underneath the turnaround is the **site of one of the few Indian ceremonial ballcourts in the Lesser Antilles**. (There are ballcourts on the Greater Antilles, starting with Puerto Rico, 60 miles to the northwest, but very few on islands further south until Trinidad.) It is believed that the Mayans exported this sport along with certain parts of their religion. Some upright stones found at this ball park site had petroglyphs. These are now in a museum in Copenhagen.

-The small rise just SE of the turnaround has remains of a triangular fort, Fort Salé, started by the English and completed by the Dutch in the early 1600s. The triangle is readily visible when the grass is low. The fort command changed between Dutch and English several times. There was also a small population of French on the island. By 1645 there were 600 colonists engaged in commerce and agriculture on St. Croix, the only settled Virgin Island. In 1650 the Spanish massacred the English on St. Croix (who had a few years earlier forced most of the Dutch and French to retreat to other islands), but later that year the French drove out the Spanish and established a colony. The center of command for St. Croix during those early years was here at Salt River and Fort Salé. (Christiansted Harbor, settled primarily by Dutch, was part of the boondocks.) The governor's residence and seat of government was somewhere on the peninsula that forms the eastern boundary of Salt River Bay, perhaps near Estate St. John according to an old French map, but by some accounts at the site of the 18th century ruins of Estate Judith's Fancy.

-Even the Indians favored Salt River. Beneath the fort, excavations have revealed the **middens of several different Indian cultures**: the Caribs, who were subjugating and replacing the pre-Tainos (often called "Arawaks") at the time when Columbus landed; the Tainos; and two other pre-Taino groups.

-You can **hike west along the beach** for a long distance, with wading required in only a few places. In about 0.75 miles, just west of the rocky point near Gentle Winds is a nice **tidepool area**. The cliffs of Baron Bluff are the first real obstacle to a beach walk, and they are about 1.1 miles away.

NORTH SHORE TOUR
INCLUDING PARTS OF ROUTES 75,80,69, AND 78 WEST
North Shore Road, Route 80

North Shore Road, Route 80
E-W (W-E)
<0.9>
6.3 (13.2) Dirt road to right or N overlooks sea cliffs. A **path goes**
2.5 (4.65) **down to sea and cobble beach.**
<0.5>
6.8 (12.7) Paved road to left or S to Windwood Estates and Estate
3.0 (4.15) Clairmont joins the Scenic Drive East, Route 78, 1 mile away.

Estate Clairmont Road:
N-S (S-N)
0.0 (1.0) Junction with North Shore Drive at mile 6.8(12.7)
0.2 (0.75) Dirt road to right or W to top of Michael Hill
0.0 Junction with Estate Clairmont Road at mile 0.2(0.75)
0.1 Parking area for two-wheel drive; last 0.1 mile may require four-wheel drive.
0.2 Turnaround on top of knoll, elevation 388 feet. This is Michael Hill, the only probable cinder cone on St. Croix. It has a depressed center (crater), and if you look at some of the rock that forms it, you will see stratified pink volcanic material. Its eroded base has formed the seacliffs called Baron Bluff.

-Michael Hill has a **scenic view** from the top. At the tip of the turnaround:

-Below in the water are a number of coral ridges. The edge of the island shelf is very close to shore here; the place where the shelf ends and changes to a steep wall is indicated by the change of water color to deep blue a little more than 100 yd. offshore.

-To the right or E: Gentle Winds Condominium; Salt River Bay; the peninsula of Estate Judith's Fancy; Green Cay; Buck Island; the peaks from east to west (far to near): twin peaks of the Goat Hills; Peak Seven of the Seven Hills; cone-shaped peak of Lang's Observatory and Lang's Peak of Carina Mountain; and flat-topped Recovery Peak.

-The **mill of Estate Clairmont**, covered and incorporated into a house on the hill above and south of the Bluff.

View from Michael Hill (continued)
-View from the W end of the turnaround:
-To the left at the seaward end of the valley south of this hill is the **mill of Estate Rust-op-twist**.
-The point to the W in the near distance is Canebay Pt.; the point in the far distance with the lighthouse on top is Ham Bluff. The high ridge spine running towards Ham Bluff is Maroon Ridge. Its highest point is 900-foot Caledonia Peak.
-Mount Eagle is the flat-topped peak east of Maroon Ridge. Its spine connects to Blue Mountain with many antennae and radio receiving dishes on top.

Estate Clairmont Road (continued):

N-S	(S-N)	
0.45	(0.55)	Dirt road to left or E goes to knoll with a view similar to the one at Michael Hill.
0.7	(0.3)	Paved road to right or W along **ruins of Estate Clairmont** (also called Estate Clearmount). The ruins of the factory and village of Estate Clairmont (along the first 0.15 mile of this road) have been given to the St. Croix Landmarks Society to form this park. They comprise an excellent example of the compact layout of a sugar factory of the 1700s. (This plantation was established in 1751.) There are plans to install plaques illustrating and identifying the various parts of the ruins. Two are easy to point out anyway. The cistern has a domed top with a small square pinnacle. The circular walled structure is the base of an animal mill or whim. Many estates had both types of mills. The animal mill was usually built first, and then served as a back-up for the more efficient wind-driven mill. The cane-grinding mill (basis for a private house) is on the hill above and S of the park.
0.85	(0.15)	Dirt road to left or E to **view of Salt River Bay and northeastern shore of St. Croix** at 0.35 mile
1.0	(0.0)	Junction with Eastern Scenic Drive [p. 127, mile 2.2(12.8)].

<0.7>

North Shore Road, Route 80 (continued)

E-W	(W-E)	
7.5	(12.0)	Dirt road shoreward to coral and cobble beach
3.7	(3.45)	

NORTH SHORE TOUR
INCLUDING PARTS OF ROUTES 75,80,69, AND 78 WEST

North Shore Road, Route 80

E-W (W-E)

<0.15>

7.65 (11.85) Dirt road to coral and cobble beach
3.85 (3.3)

<0.15>

7.8 (11.7) The **mill of Estate Rust-op-twist** is on the inland hill.
4.0 (3.15) "Rust-op-twist" is Dutch for "rest after toil".

<0.1>

7.9 (11.6) Inland are the private **old buildings of Estate Rust-op-twist.**
4.1 (3.05) -Route 80 between this and the next road is lined with **Coconut Palms** and **Otaheita**. Otaheita, a member of the Mallow or Hibiscus Family, has attractive yellow flowers and is widely distributed in the tropics along shorelines.

<0.1>

8.0 (11.5) Dirt road inland to Rust-op-twist residential area. You can
4.2 (2.95) drive a loop here around the Rust-op-twist Marsh, a **good birdwatching area** after you penetrate the bush.

Loop around Rust-op-twist Marsh (E-W):

0.0 Junction with Route 80 at 8.0(11.5). At 50 yards further make a right turn to S.
0.1 Junction with dirt road uphill to left. Go right.
0.45 Turn right or W.
0.7 Turn right or N.
0.95 Junction with Route 80 at mile 8.2(11.3)

<0.05>

8.05- (11.35- Beach access on several dirt roads. Beach is mixed cobble
8.15 11.45) and sand. **Manchineel Trees** provide welcome shade.
4.25- (2.8- However, one must be wary of these trees. Their little
4.35 2.9) green apples are poisonous, and some people are allergic to the sap and leaf coatings. People who stand under them during rain sometimes develop a rash.

<0.6>

8.65 (10.85) Junction with Route 73 at village of LaVallee (pronounced
4.85 (2.3) "La Valley"). To continue Route 80 and the North Shore Drive, turn right or W. [See "Side Trip Along Canaan Ridge Road", p. 131, mile 0.0(4.7) for the log of Route 73. This road winds through moist forest and connects with the Scenic Drive (1.15 miles), Midland Road (3.3 miles), and Centerline (4.7 miles).]
-The Sea Shadows **SCUBA Dive Shop** is on the SW corner of this intersection, telephone 778-3850.
-**West Indian Mahogany Trees** line the road ahead.

<0.2>

8.85 (10.65) Route 80 swings right or N. A dirt road goes straight or W,
5.05 (2.1) passing inhabited **ruins and mill of Estate LaVallee.**

<0.2>

9.05 (10.45) Road swings to left or W along the shore.
5.25 (1.9)

<0.05>

NORTH SHORE TOUR
INCLUDING PARTS OF ROUTES 75,80,69, AND 78 WEST
North Shore Road, Route 80

E-W (W-E)

9.1 (10.4)
5.3 (1.85) Grassy path to shore, opposite E side of school yard. The white sand beach has some cobbles. **Tidepools** are in beach rock, a former beach shelf or reef with many of the coral skeletons fossilized in the rock. This beach rock erodes to a very sharp-pointed irregular surface, called *diente del perro* or "dog teeth." Sometimes areas erode underneath and through the top of the shelf, creating blowholes. One blowhole is on the path to shore. It is usually blocked with sand except during winter storms. One can walk the beach and beach rock for some distance in both directions.

-About 25 yards W of the path, opposite the house with the blue roof, is a view of the **ruins, mill, and chimney of Estate LaVallee** inland.

<0.6>

9.7 (9.8)
5.9 (1.25) Ahead to W is **Mount Eagle** (1165 feet), the highest point on St. Croix.

-Two dirt roads lead inland. The left (Vallee View Road) goes through residential areas with **views of the north shore**, ending in 1.25 miles at the Scenic Drive [p. 130, mile 4.35(10.65)]. The right goes 0.3 mile to dead end.

9.7- (9.6-
9.9 9.8)
5.9- (1.05-
6.1 1.25) **Dive and beach access to Cane Bay coral gardens and wall.** The coral reef at Cane Bay is one of the highlight dives on the island. The best entry point is at the boat ramp ahead. On the seaward side of the entry area is a white-fenced beach with many coconut palms. The corals offshore are arranged in the classic groove and buttress pattern of interdigitating sand trenches (grooves) with ridges of live coral topping mounds of coral rubble (buttresses). The slope changes dramatically at 70 feet deep (150 yards offshore) from 5% to 75%. To make a dive, snorkel straight out N until you are just losing sight of the bottom, then descend where you can still see the bottom. With usual visibility conditions you should hit the bottom at 60-70 feet deep. If you want to make a deeper dive, work north along the bottom and you will quickly reach the edge of the St. Croix shelf, which plunges steeply to over 10,000 feet deep. High drama is here, as you float amidst deepwater fishes and tendrils of black coral, watching the sea floor disappear into the abyss.

<0.1>

9.8 (9.7)
6.0 (1.15) Paved road inland to Cane Bay Plantation Hotel amid **ruins of Estate Cane Bay.**

<0.1>

9.9 (9.6)
6.1 (1.05) Boat ramp and good entry point for Cane Bay dive.

NORTH SHORE TOUR
INCLUDING PARTS OF ROUTES 75,80,69, AND 78 WEST
North Shore Road, Route 80

E-W (W-E)

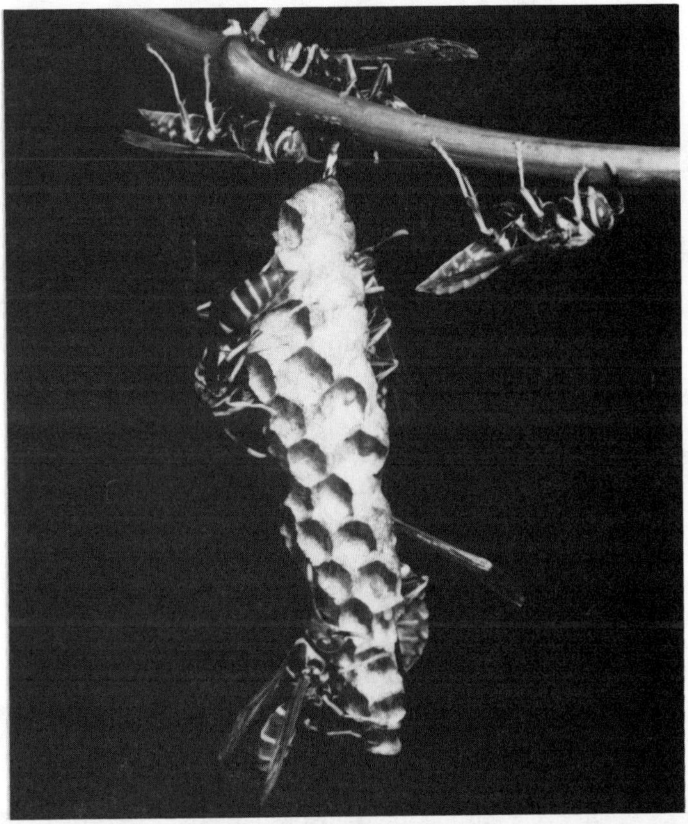

Fig. 37. These Jack Spaniards readily inflict a very painful sting when defending their nest. Anyone brushing the nest, which often hangs hidden in brush, is considered an attacker. These wasps are surely one of the chief hazards of off-road hiking.

<0.1>

10.0 (9.5)
6.2 (0.95)

Dirt road inland. **Ruins of Estate Cane Bay** are on the SW corner of junction. This is a very fine set of sugar factory ruins. You must hike a short distance through light bush, watching for wasp nests, to get to the factory. It has many old lintel timbers still in place. Also, the only ruins on St. Croix with all three coppers (iron boiling pots for sugar juice) in place is this factory. One has the draining tiles still in place. You can look under the coppers at the fire pit areas and ash removal channels. A short walk from the south end of the factory is the circular wall of the animal mill. Just south of the animal mill is the cane-grinding windmill, built in 1820. Lift some of the timbers lying on the mill floor. Termites would have demolished

E-W (W-E)

lighter wood long ago. It takes them a long time to consume dense wood such as Lignumvitae or Bulletwood, which may be the material in these pieces. This is one of the few mills with a fireplace and flue in the wall, the flue now blocked by a bees nest. The fire might have been for on-site blacksmithing and heating meals.

<0.1>

10.1-10.2 (9.3-9.4) 6.3-6.4 (0.75-0.85) **Turnouts to beach**

<0.25>

10.45 (9.05) 6.65 (0.5) Private house built into **mill and ruins of Estate North Star** on inland side.

<0.2>

10.65 (8.85) 6.85 (0.3) Beach access, just before (E of) road swings inland. This is access point for North Star **wall dive**, where the first part of the wall starts at 40-50 feet. For this dive, entry is the same as at Cane Bay: snorkel out until bottom corals seem to disappear. Descending where you can still see the bottom will put you at 40-60 feet.

<0.15>

10.8-10.9 (8.6-8.7) 7.0-7.1 (0.05-0.15) Road dips inland through a pocket of moist forest with large trees and *Cissus* vines. Roots of the vines trailing to the ground give a dense aspect to the forest. The roots of these vines grow downward from the tops of trees. Birds eliminate the seeds high in the trees; they germinate and grow without a need for soil.

<0.05>

10.85 (8.65) 7.05 (0.1) Dirt road to left or S just where road begins to swing right or N again. This road is an access point for a pleasant **hike through the shady northwest forest of St. Croix**. There is no actual trail. It is best for those in passenger cars to park on the opposite side of the paved road and walk in along the dirt road. Two very large trees on the opposite side of the dirt road are **Mammee Apples**. They have erect trunks, large, shiny leaves, and edible brown fruits 3-10 inches in diameter. The dirt road ends at 0.05 mile near a metal pipe well. Go right or S from this pipe well, through the open forest about 100 feet to the usually dry creek bed. Follow the creek bed upstream about 100 feet and take the right fork. Follow this several hundred feet and take the left fork. The hike from here on is a climb up the creek bed. Throughout the walk you may see the ground littered with hard seed pods, about 2 X 5 inches and 0.5 inches thick. These are from a native plant that is common in the moister forests of St. Croix, the **West Indian Locust**. The seed pods are common drift seeds, washing up on shores and then sprouting. You may

E-W (W-E)

hear calls of a bird of the moist forests, the **Bridled Quail-Dove**. This bird gives a mournful, low "whooo", that sounds like someone blowing over the mouth of a bottle. The Quail-Doves nest on the ground, which makes them susceptible to mongoose predation and contributes to their rarity.

-About 30 yards beyond the road above, a dirt road to right or N leads to **beach access for a white sand beach**, at 0.05 mile. One can beachcomb some distance both E and W.

<0.1>

10.95 (8.55) Junction with River Road, Route 69 [p. 133, mile 0.0(4.4)].
7.15 (0.0) Route 80 ends here. Route 69 connects with Mahogany Road (3.5 miles) and Centerline (4.4 miles). To continue the North Shore Drive, turn left or S onto Route 69.

-The dirt road that continues straight or W along the shore goes to Davis Beach, a **white sand beach with access for wall dive**. This area is being developed into the Carambola Beach Resort. Inquire at the hotel regarding beach and ruin access. On the property are **ruins of Estate Prosperity** including the old **greathouse at the beach**; a **chimney** with a nearby **factory**; and a picturesque, split, **cane-grinding windmill**. The access for a **wall dive** and a **hike to Annaly Bay** are described below.

-The stone ruins of **Estate Prosperity greathouse** are just inland from Davis Beach. The greathouse is unusual because it is made of dark Caledonia rock. Most other greathouses were constructed of coral blocks and quarried limestone marl.

-Davis Beach is **a scenic white sand bay amid coconut palms, with towering cliffs to the west**.

-To make a **wall dive** here, the procedure is the same as at Cane Bay. Snorkel out until you begin to lose sight of the bottom, then descend. You will be at about 60 feet.

-One can **hike the system of paths that leads to Annaly Bay** west of Davis Bay, but it takes some scrambling at the start to reach the path. First climb the rocky wall that is the west border of Davis Beach. (You are now on the headland pictured in the last scene of the movie *Trading Places*.) Then wend uphill through bush to grass. Climb up along the ridgeline of the grassy slope to a slight break in the slope. A clump of trees is on this more level area, which is about 100 feet in elevation. At 200 feet elevation, you will come to a fairly well-worn cattle trail that leads off to the right or W along a level contour to the start of a trail (formerly the road to Estate Sweetbottom in the 18th and 19th centuries). (If you go too far up, at about 100 feet further distance you will encounter the beginning of the forest. A clump of trees with very stiff, sandpapery leaves that crackle when crumpled is on the

Davis Beach (continued)
ridgeline at this point. Go back along the western side of the grass-forest edge, and you should encounter the path.) [The system of trails around Furnel Ridge is described with a map on p. 117, mile 13.2(6.3). The map shows the connection to the bluff at Davis Beach.] Once you surmount the bluff at Davis Beach, the rest is easy, for elevation changes are very gradual from this point on.

-You can go rock-hopping along the shore to Wells Bay (on the E of Furnel Point, the one seen in the distance from Davis Beach). This route requires periodic wading. You will probably get wet up to your waist at an area about 0.1 mile W.

<0.75>

River Road, Route 69

E-W (W-E)

11.7 (7.8) Junction with Scenic Drive West, Route 78, [p. 131, mile
0.75 (3.65) 7.2(7.8)]. Turn right or W onto the dirt road of the Scenic Drive to continue the North Shore Tour.

-River Road continues straight or S [p. 133, mile 0.75(3.65)] through a rural area, 1.25 miles to Route 72, and 4.4 miles to Centerline. The Scenic Drive to the left or E winds through the highest forest on the island with viewpoints of both coasts. [The first 7.2 miles are in "Side Trip on Scenic Drive", p. 131, mile 7.2(7.8).]

-One can stand at 560 feet elevation on the knoll on the NE corner of this junction for a **nice view of the north shore of St. Croix and the other Virgin Islands to the N on a clear day**. Clear days are rare from May to September when high level winds transport 30 million tons of red-brown dust from the Sahara of Africa to the Caribbean. In the years since measurement began in 1965, the amount of dust has increased more than three-fold, probably due to Sahara droughts coupled with land abuse. No man (or island) is an island. Africa loses its precious topsoil, and St. Croix looks like it has smog.

-Just E of this point is the summit of flat-topped Mount Eagle, the highest point on the island (1165 feet).

-The dirt road (no vehicles allowed) on the knoll on the NW corner of the junction climbs uphill to Bennykenny Hill, 713 feet, and then descends steeply downhill (20% grade) through north-slope forest 0.6 mile to Davis Beach. It gives a **good view of both sides of the island**.

We have listed some points of interest off the Scenic Road West in the next portion of this tour west of Route 69. Much of that area is a private cattle ranch, which at this time, has "No Trespassing" signs on only a few roads. If you wish to go beyond a gate, walk--do not drive in. Please leave gates as you find them. Any hiking you do in this area is at your own risk.

NORTH SHORE TOUR
INCLUDING PARTS OF ROUTES 75,80,69, AND 78 WEST
Scenic Drive West, Route 78

E-W (W-E)
<0.3>
Scenic Drive West, Route 78
E-W (W-E)

12.0 (7.5) **Fountain Valley overlook.** Sweeping left to right or E-W:
7.5 (7.5) -**Mill of Estate River** is on a conspicuous grassy hill below.
-The mill is almost in line with **Ruth Island**, the dredge island offshore from St. Croix's biggest **mangrove, Krause Lagoon.**
-Hess Oil Refinery on the south shore.
-The tall, **round chimney at Bethlehem Old Works Sugar Factory** and the shorter, **rectangular chimney at Lower Love** Agricultural Station are south of the River Mill.
-The bright red-roofed buildings of University of the Virgin Islands are further south between these two chimneys.
-The **mill of Estate Big Fountain** is on a low hill on the western side of Fountain Valley, very near the clubhouse of Carambola Golf Course.
-On the skyline to the W is the **mill of Estate Bodkin.**

<0.8>
12.8 (6.7) Concrete section of scenic drive begins.
8.3 (6.7)

<0.05>
12.85 (6.65) Barbed wire gate to left or S marks end of walk to east
8.35 (6.65) described at mile 13.5(6.0) [p. 119].

<0.15>
13.0 (6.5) End of concrete section at green mileage marker "6".
8.5 (6.5)

<0.1>
13.1 (6.4) A view of northern shore is here. See mile 13.15(6.35).
8.6 (6.4)

<0.05>
13.15 (6.35) **View of south and north shores of St. Croix from 830 feet**
8.65 (6.35) **elevation.** Sweeping right to left, S-N:
-Red roofs of the University of the Virgin Islands
-Ruth Island offshore from Krause Lagoon
-Round tanks of former Martin Marietta Alumina Plant
-Inland, the chimney at Bethlehem Old Works Sugar Factory
-Cracking towers at Hess Oil Refinery
-Watchho, a low shoreline hill to left of the oil refinery
-The mill of Estate Diamond Keturah on a low hill to E of Watchho
-Flat-topped Recovery Hill
-Blue Mountain (1096 feet) with numerous antennae and radio dishes
-Mount Eagle (1165 feet), flat-topped, the highest point on St. Croix
-Seaward tip of land is Canebay Pt.

NORTH SHORE TOUR
INCLUDING PARTS OF ROUTES 75,80,69, AND 78 WEST
Scenic Drive West, Route 78

E-W (W-E)

Coming towards you, E-W:
-Red roofs of Cane Bay Hotel group
-**Mill of Estate North Star**, incorporated into a residence
-**Mill of Estate Prosperity**, split in two
-Carambola Beach Resort at Davis Beach below you.
-The **interdigitating groove and buttress system of the coral reef of Davis Bay** is readily visible on a sunny midday. Polarized sunglasses help you see into the water. The vivid turquoise color indicates the sand bottom of the grooves; the deep green indicates the hummocks or buttresses of coral reef. The abrupt end of the buttress and grooves and deep blue color indicates the shelf edge where the slope steepens from about 5% to 75% and plummets to the ocean floor over 2 miles deep.
-On a clear day you can see the **northern Virgin Islands** 40-60 miles away. From left to right, W to E: St. Thomas, St. John, Tortola, Virgin Gorda. You actually see peaks of their hills so it appears as though there are more than four islands.
-On top of the cliffs of the roadcut behind you is a **mountaintop moist forest** typical of St. Croix. Using the vegetation for support, climb to the open grassy area on top starting at the red-painted rocks. A partially overgrown trail leads south downhill and slightly left. In less than 200 feet the trail turns slightly right or W and leads downhill about 0.3 mile to mile 13.5(6.0) [p. 119].

<0.05>

13.2 (6.3) Pullout to right or N for **view of the northwestern coast**.
8.7 (6.3) The lighthouse on Ham Bluff is at 360 feet on the western end of the spine called Maroon Ridge. The highest point is the peak of the ridge nearest you called Ham Peak of Caledonia Mountain (916 feet). On a crystal clear day you can see the peak El Toro of the Luquillo Forest on Puerto Rico, 3524 feet high, 65 miles away.
-This viewpoint is also the trailhead for a **hike along Furnel Ridge to the north shore**. The trail is fairly steep: it drops from here (about 640 feet) to the shore in a distance of about 0.4 mile (30% grade). The trail is overgrown for most of its length. Wear hiking shoes and long pants, and carry water. The path starts off in a grassy area, nearly obscured by grass and brush, becoming more distinct as you walk down the ridge line in WNW direction, heading towards Hams Bluff and the lighthouse. Try to stay on the ridge, keeping both sides of the path sloping away from you, and you will be able to pick up the trail again if you lose it.

Furnel Ridge Hike (continued)

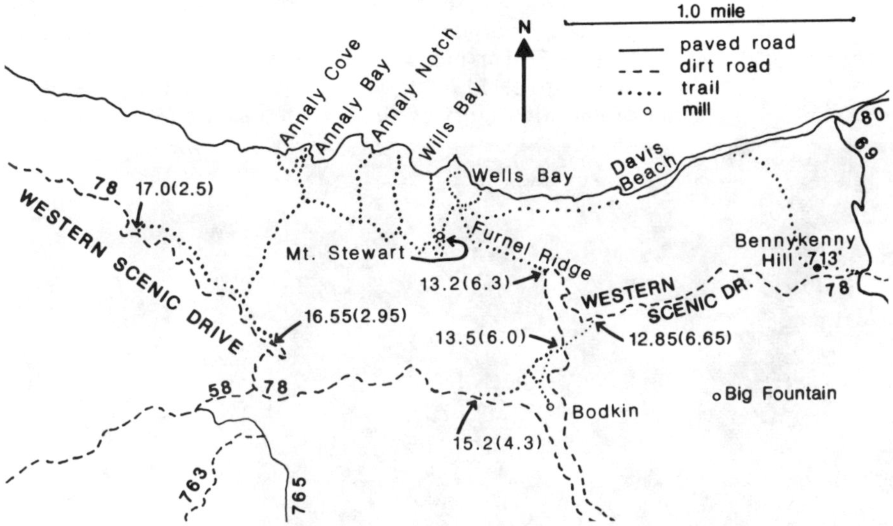

Fig. 38. Map of hiking areas near Furnel Ridge and Annaly Bay.

The walk leads through **moist forest of native vegetation. Anthuriums** called "Wild Tanier" (pronounced "tan-yah") are abundant. A native **Tyre Palm**, extremely rare on St. Croix, is also near the trail on the left or S side, and others can be seen in the distance along its length. The path comes out into an open grassy area at 200 feet elevation. From there you can see large platform rocks in the water near the shore, often with fishermen on top who have walked along the shore from Davis Beach. This is the western side of Wells Bay, ending in Furnel Point.

-From the low point of the saddle of the grassy slope, a 200-year-old roadbed leads E back into the forest, about 20 feet lower than where you emerged onto the grassy area. This is the old road to Sweetbottom Estate, which emerges on the bluff east of here over Davis Beach. The old road, now an indistinct trail, stays fairly level at 160-180 feet elevation. The path is completely in shady forest, about 0.6 mile. Along the way you can see **Wild Tanier** and **Strap Ferns**. The hike up the bluff is described under Davis Beach [p. 113].

-You can descend to the platform rocks of Wells Bay by way of the grassy saddle or along the forest edge, but both ways are scrambles and brushy.

-The old roadbed leads inland or S downhill from the grassy saddle. You can follow this road to get to **more bays with sandy beaches** on the north shore, as well as the **mill and ruins of Estate Wills Bay** (included in Estate Mt.

Furnel Ridge Hike (continued)
Stewart). You will encounter many webs of the Golden Silk Orb Weaver Spider. Carry a stick to clear them before you. [See p. 76 for more about the spiders.]
-The map on p. 117. shows the various sites. Here are some of the highlights of this isolated section of shoreline.

Fig. 39. Wild Tanier (Anthurium) is abundant on the hike down Furnel Ridge.

-Hiking out to the western edge of the grassy saddle of Furnel Ridge gives a good view of the western side of Wills Bay, the next one west of Wells Bay (Confusing? See the diagram.) On the western slope of Wills Bay, there is a dramatic change of the seaward windsheared forest, unsheltered from the Trade Winds, to the inland upright forest in the lee of Furnel Ridge.
-Wills Bay is reached from an offshoot of the old roadbed. It has a lovely, deserted cobble beach. From this point you get a good look E at Furnel Ridge as it descends to Furnel Point. The **tidepools** on the NW point are quiet on most days. Some are about 6 feet deep. An old reef deposited on the black Caledonia bed rock is uplifted to form a very large seaward barrier for the pools. It is fantastically eroded to a jagged nightmare. With some cliff scrambling, one can walk all the way between Wills Bay and the cove west of Annaly Bay and view the **tidepools** on the way.

Furnel Ridge Hike (continued)

-At a fork, turn left or N to the **Estate Wills Bay mill** (included in Estate Mt. Stewart).

-At the point where the road climbs out of the shady forest, an old overgrown road takes off to the north. This road used to go to Annaly Bay Notch but now is overgrown in the final part. If you walk along the nearby stream bed (or paths which run along the banks) that are just east of this junction, an easy walk through a lovely shady forest will take you to **Annaly Bay Notch**. Some of the best tide pools are nearby.

-**Annaly Bay has beautiful tidepools** between the NE side and the Annaly Bay Notch.

Fig. 40. Golden Silk Orb Weaver Spiders are in all the forests.

<0.3>

Scenic Drive West, Route 78 (continued)

E-W (W-E)

13.5 (6.0) Just S of the green mileage sign "5.5" are dirt roads to E and
9.0 (6.0) W behind barbed wire gates. Leave the gates to Scenic Road closed, regardless of how you find them. If you want to see the area described here, leave your vehicle on the Scenic Drive and walk. The road to the right or W goes 0.3 mile, to rejoin the Scenic Drive at mile 15.2(4.3). Walk in off the Scenic Drive and both the first and second roads you come to on the left lead to **Bodkin Mill**. (Total walk to Bodkin Mill is about 0.4 mile one-way.) The best view is on the hill just NW of the mill. It's a **spectacular**

E-W (W-E)

panorama of St Croix, almost 360° from 960 feet elevation. [See mile 13.85(5.65) below, for a description of the view.] If you know exactly where to look, you can see 13 mills from this point. The Bodkin Mill, highest on the island, is secluded and scenic. Note the 1808 plaque over the entry. Sugar cane was unloaded at this opening. After the cane was pressed through the rollers, the bagasse (crushed stalks) was passed out the smaller opening to the north, dried, and then used as fuel to boil down the cane juice. The cane juice flowed out the gutter exit (smallest opening) downhill to the south. The opening opposite the entry is an uncommon, luxurious feature: a window to let in the trade winds. If you had to be a slave in the millworks, you might have preferred Bodkin. What is the tall narrow slit for? Think like a mechanic, then guess. This is the exchange slit, for getting long machinery parts, especially the shaft, in and out of the mill.

-The road to the left or E goes 0.35 mile to mile 12.85(6.65) [p. 115] of the Scenic Road above. It has a junction with a dirt road to the N which climbs up to the top of Furnel Ridge through **native moist forest**.

<0.15>

13.65 (5.85) **Mill of Estate Bodkin** is on a hill to the right or W. A
9.15 (5.85) **grove of Teak Trees** is on the hill to left or E.

<0.2>

13.85 (5.65) **Viewpoint for both south and north shores from 880 feet.**
9.35 (5.65) Sweeping from left to right, N-S, then E-W:

-Eagle Ridge, with flat-topped Mount Eagle (1165 feet) nearby and Blue Mountain (1065 feet) with numerous antennae at its eastern end

-Fountain Valley is in the foreground, with **mill of Estate Parasol** in a saddle between two low hills between Fountain Valley and Blue Mountain.

-Goat Hills, two cone-shaped peaks in far distance near eastern end of St. Croix

-Peak Seven of the Seven Hills Range

-Flat-topped Recovery Peak. Christiansted is hidden at its foot.

-Watchho (also Vagthus Point) is the point on the south shore east of Hess Oil Refinery. The **mill of Diamond Keturah** is on a low rise to the east of Watchho.

-Seaward of the former alumina plant are Krause Lagoon and Ruth Island.

Then coming inland towards you:

-**Mill of Estate River** is on the SE side of Fountain Valley, on a low grassy hill in line with Hess Oil.

Fig. 41. Interior of Bodkin Mill

-The **chimney of the Bethlehem Old Works Sugar Refinery** is the tall, round chimney on the plains below, almost mid-island. A **square chimney of Estate Lower Love** is to the right or W on the grounds of the Virgin Islands Agricultural Station.

-**Ruins** of at least 11 estates are visible from this point.

<0.15>

14.0 (5.5) **View** as above

9.5 (5.5)

<0.1>

NORTH SHORE TOUR
INCLUDING PARTS OF ROUTES 75,80,69, AND 78 WEST
Scenic Drive West, Route 78

E-W (W-E)

14.1 (5.4) Highest point of road at 950 feet (no view) just S of green
9.6 (5.4) mileage sign "5"

<0.2>

14.3 (5.2) **View of the southwestern coast**, with features as mentioned
9.8 (5.2) at 13.85(5.65) above

<0.1>

14.4 (5.1) **View of the southwestern coast** at southern point of hairpin
9.9 (5.1) turn. In addition to shoreline features mentioned at 13.85(5.65) above, one can see the **chimney of Estate Two Friends, mill and ruins of Estate Orangegrove** on a low hill, and St. George Hill in the distance with numerous antennae on top and a rock face produced by quarry operations. One can also see Frederiksted and its pier.

<0.1>

14.5 (5.0) **View of southwestern shore** as at 14.4(5.1)
10.0 (5.0)

<1.65>

16.15 (3.35) Junction with Creque Dam Road, Route 58, at green
11.65 (3.35) mileage sign "10". To continue the North Shore Tour and the Scenic Road, veer right or N. The left fork to W is Creque Dam Road, which goes 3.2 miles through "rain forest" to join Route 63. [See "Side Trip Along Creque Dam Road," p. 92, mile 3.2(0.0).]

-After heavy rains, the final mile of the Scenic Road is sometimes four-wheel drive only. Creque Dam Road is usually in better condition. On Creque Dam Road just 0.25 mile from this junction is paved road Route 765, which can be used for a quick exit from the Scenic Drive.

<0.4>

16.55 (2.95) Dirt roads to Annaly Bay are on the right or N side of a
12.05 (2.95) barbed wire fence here. To see this area, park on the Scenic Drive and then walk. [See map, p. 117.]

Roads of the Annaly Area (S-N):

0.0 Fenceline. Leave this gate closed regardless of how you find it. The left fork runs 0.5 mile to 17.0(2.5) [p. 123] overlook of the bays along this part of the north shore. The right fork goes downhill to Annaly, Wills, and Wells Bays, and **ruins of Estate Wills Bay** (included in Estate Mt. Stewart). Take the right fork.

0.5 Bypass loop to right or E, which reenters main road about 0.05 mile further

0.7 Turn right or E for roads to Wills and Wells Bays, **ruins of Estate Wills Bay**

Road to Estate Wills Bay (W-E):

0.0 Junction

0.1 Dirt road to left or N goes to Annaly Bay Notch with a gravel beach, 0.2 mile distant.

Roads of Annaly Area, S-N (continued):

Road to Estate Wills Bay, W-E (continued):

This road is very indistinct towards end. The best route to the beach is along the stream bed just passed before this road. Annaly Bay Notch has some very nice **tidepools**.

0.2 Young **grove of West Indian Mahogany Trees** on the left or N

0.55 Three-way fork. On the NW corner of this junction are the **ruins of Estate Wills Bay**. The road to extreme left runs 0.2 mile to Annaly Notch with a gravel beach. The middle road goes to the **mill of Estate Wills Bay**. This road ends at 0.25 mile on a bluff between Wills Bay and Annaly Notch. To continue take the right-most fork to E.

0.6 Fork. The left fork goes to Wills Bay (also called Stewart Bay) at 0.2 mile distance, with a gravel beach and an impressive elevated fossil reef exposed high on the underlying black Caledonia formation. The right or E fork goes to the grassy saddle of Furnel Ridge overlooking Wells Bay and mentioned in the hike description on p. 116, mile 13.2(6.3).

Annaly Area, S-N (continued)

0.85 Fork. The right or E fork goes about 0.05 mile to bluff overlook of Annaly Bay. The left fork goes W 0.15 mile to a cove on the west side of **Annaly Bay**. Walk through the trees to **sand and rock beaches**. Nice **tidepools** are formed in the beachrock shelf.

<0.45>

Scenic Drive West, Route 78 (continued)

E-W (W-E)

17.0 (2.5) **Viewpoint for Annaly, Wills, Wells, and Cane Bays** going
12.5 (2.5) from left to right, W-E. The western end of Davis Beach is hidden. Canebay Pt. forms the eastern boundary of Cane Bay. A dirt road leads east here to a barbed wire gate at about 0.1 mile to join the Scenic Drive 0.5 mile away at mile 16.55(2.95)

<0.3>

17.3 (2.2) **Mill of Estate Springgarden** is on a hill across Caledonia
12.8 (2.2) Gut on the left or S side of road. Roadside brush is often dense, partially obscuring the view.

<0.85>

18.15 (1.35) Dirt road to right or N goes 0.4 mile to end around the base
13.65 (1.35) of **Scratch Peak**. The road may be posted "No Trespassing", and you will need to ask permission to enter. The road is bad. Do not drive in; park on the Scenic Drive. The road ends at the head of a windswept, grassy gully, which has **native West Indian Spider Lilies**. The

E-W (W-E)

lilies are usually in bloom August-September. From the end of the road, walk onto the top of the grassy ridge. Walk through grass along the ridge line uphill along an indistinct trail. At the end of the grass, an indistinct path leads into woods. The path follows the ridgeline, through a **moist forest** with some **Strap Ferns**. The peak of the trail opens onto a grassy patch which affords a **view to the northeast along the shoreline** of Annaly, Wills, Wells, Davis, and Cane Bays. Canebay Pt. and the cone of Michael Hill are in the eastern distance. About 200 feet east of this overlook one can see rock outcrops which are covered with **Christmas Orchids**. Walk down the grassy slope to pick up the path into the woods, and follow it to rock outcrops and orchids. Take pictures, not orchids. The total distance from the road's end is about 0.25 mile.

<0.2>

18.35 (1.15) Dirt road to right or N to top of Horse Peak at 640 feet.
13.85 (1.15) The road may be posted "No Trespassing", and you will have to ask permission to enter. Walk up the embankment; do not drive. A nice **view** of the lighthouse on Ham Bluff and (on a clear day) Puerto Rico and the northern Virgin Islands: St. Thomas, St. John, Tortola, Virgin Gorda (from left to right or W-E). To the south across the Caledonia Valley, a gorge about 450 feet deep at this point, one can see the **mills of Estates Mount Victory and Nicholas** (left and right). The very dark rock face of the **old quarry works** near the mouth of the Caledonia Valley gave these rocks their geological name, Caledonia Formation.

-A very steep, concrete-paved section starts here. The section of the road which washes out in heavy rains is at the end of the concrete, and two-wheel drive vehicles may have to return uphill at that point. An easy exit is available at mile 16.15(3.35) above.

<0.3>

18.65 (0.85) End of concrete section of road.
14.15 (0.85)

<0.45>

19.1 (0.4) **Mill of Estate Ham Bay** to right or W of road.
14.6 (0.4)

<0.25>

19.35 (0.15) Pavement begins. To continue the tour make a right or W
14.85 (0.15) turn here. To the left or E is a road (sometimes fenced) which leads into the Caledonia Valley, the only gut with year-round flowing water. The entry road is privately owned. The easement for entry to the 95 acres owned by the V.I. Government (to be a nature preserve) is the gut itself, so walk in the streambed to enter. The **quarry in the Caledonia Formation**, now closed, has an impressive 300-foot high exposure on the wall of the valley. The

E-W (W-E)

dark metamorphosed rock was deposited 80 million years ago on a deep sea floor, eroded material from a volcanic source. A **beautiful hike** of about 1 mile continues up **Caledonia Gut**. Rock pools contain a variety of native freshwater fish and crustacea. The lush growth along this watercourse includes some horticultural introductions. **Vanilla Orchids** (the same species from which true vanilla flavoring is made) festoon some of the trees. Just after heavy rains there is a waterfall about 50 feet high at a confluence about 0.5 mile up the gut from the quarry.

<0.15>

19.5 (0.0)
15.0 (0.0) Junction with West End Road, Route 63 [p. 79, mile 6.15(0.8)]. Two stone pillars here flank a turn to the right or N which leads to Ham Bluff.

Note: If you are driving this tour (W-E), in less than a mile you will reach a short section that may require four-wheel drive. If it is not passable, start this tour from the junction of Creque Dam Road (Route 58) and Route 63 [p. 88, mile 0.0(3.2)] or from the junction of Mahogany Road (Route 76) and Route 765 [p. 83, mile 3.25(6.4)].

Fig. 42. Caledonia Gut, a territorial park filled with shady rock pools.

Fig. 43. Bromeliads decorate large trees on Creque Dam Road and in the other moist forests.

SIDE TRIP ON SCENIC DRIVE, ROUTE 78

This drive runs along the spine of the northwestern mountains of St. Croix, and affords wonderful views of the lowlands along both north and south shores. The tour starts on "Scenic Drive East" off of the North Shore Drive, Route 80. Scenic Drive East goes 3.8 miles to Canaan Ridge Road, Route 73. One then must go 0.15 mile N on Route 73 to the start of Scenic Drive West. Scenic Drive West then goes 3.8 miles to River Road, Route 69. This section is on the highest mountain of St. Croix, Eagle Ridge. The most popular section is the final run after it crosses River Road, Route 69, and that segment is discussed as part of the North Shore Tour, starting on [p. 114, mile 11.7(7.8)].

Scenic Drive East, Route 78
E-W (W-E)

0.0 (15.0) Junction with Route 80 [p. 102, mile 4.15(15.35)].

<0.5>

0.5 (14.5) Pullout on right or E of a curve. View of watermill, chimney, and cane-grinding mill of Estate Morningstar across the valley.

<0.2>

0.7 (14.3) End of pavement

<0.1>

0.8 (14.2) Fork. Go right or NE.

<0.85>

1.65 (13.35) **View of Salt River Bay** at the wooden sign "Concordia Ridge Road". One can see Salt River Estuary below, Christiansted Harbor, Green Cay, and Buck Island in the distance. The two cone-shaped hills at the far end of the island are the Goat Hills. The higher peak is the northern one, Goat Peak (672 feet).

<0.1>

1.75 (13.25) At the driveway of plot 66A is a good **view of the south shore industrial complex** of Hess Oil Refinery and now-closed Martin Marietta Aluminum Refinery. Because of the incline of the road at this point, park your car in the flatter area ahead opposite the road marked "Do Not Enter," and walk back for the view.

<0.20>

1.95 (13.05) Driveway to right or E to view at plot 88, a vacant lot. A **nice view of Salt River Bay and the northern coast of St. Croix to the east**. See description of view for Windwood Estates Loop below.

<0.25>

2.2 (12.8) Three-way fork. To continue the Scenic Drive, take the leftmost or southern fork and stay at about the same elevation.
-The middle fork to N heads downhill to Estate Clairmont and some views of the north shore, connecting after 1.0 mile with Route 80 [p. 107, mile 6.8(12.7)].
-The rightmost fork to E heads uphill. This is a one-way loop road through Windwood Estates and affords some

very nice **views of Salt River Bay and the northern shore of St. Croix** to the east.

Windwood Estates Loop (one-way W-E):

0.0 Junction with Route 78 at mile 2.2(12.8)

0.05 Three-way fork. Take the leftmost road. The house ahead of you is on Clairmont Peak (866 feet) of Saltriver Mountain.

0.2 **Viewpoint eastward from this hill** at plot #99, just before a steep downhill section of road. On the plain below and moving away from you to the E:

-Gentle Winds Condominium with **mill of Estate Salt River** inland S of it

-Mouth of **Salt River Bay**. The mouth or opening in the reef which protects the bay is at the inland edge of a submarine canyon, an underwater topographic feature that is much like a narrow river gorge. You can detect its presense on a sunny day by the deep blue color of the water over the canyon, bordered on both sides by greenish-colored water. The canyon appears to trend northwest from this vantage point, pointing as a finger into Salt River Bay.

-Inshore is the white sandy beach with a paved turnaround just inland of the beach. This is the site called "**Columbus' Landing**". [See p. 104, mile 5.4(14.1)].

-The **Hemer Peninsula** forms the eastern boundary of Salt River Bay and is called Estate Judith's Fancy. About 0.25 mile north offshore of the point is **a shoal area** that usually has breakers even in fairly calm weather, **White Horse**. The 1977 **wreck** of the island freighter *Cumulus* is on Salt River Reef, just west of the point.

-The windmill, chimney, and ruins of Estate St. John are in line with Christiansted. **Long Reef is the coral barrier reef** that forms Christiansted Harbor. **Protestant Cay** is the island in the harbor.

East of Christiansted:

-The radio tower of WSTX is on the hill of a point where **Fort Louise Augusta** once stood. The radio station is built amid the ruins of the old Fort. The large **mangrove** just inland of the radio tower is **Altona Lagoon**. A light turquoise area in the water north of the Fort is the seaward border of **Scotch Bank**.

-The first island along the shoreline to the east is **Green Cay**, a National Wildlife Refuge for an endangered species of lizard. [See p. 18.] In line with the isthmus of Green Cay is the **sandbar of**

Side Trip on Scenic Drive, Route 78

Scenic Drive East, Route 78

Windwood Estates Loop, one-way W-E (continued)

Candle Reef. Further east is the **barrier coral reef of Tague Bay**. The Bay is light greenish blue in contrast to the royal blue of **Buck Island Channel** offshore, which lies between Tague Bay Reef and Buck Island.

-The two peaks in the far distance are the **Goat Hills Range**. The most northern peak is the higher one, Goat Peak (672 feet).

Moving inland and going S:

-The large mountain west of Goat Hills is the **Seven Hills Range**. The high point is Peak Seven, 866 feet.

-The next mountain to the west is called **Carina Mountain**. It has a low but prominant peak, called Lang's Observatory (425 feet), then a flat-topped peak with numerous antennae and receiving dishes (visible with binoculars), and then another cone-shaped peak, Lang's Peak (726 feet).

-**Recovery Peak** (845 feet) of Recovery Hill is distinguished by its broadly flattened top with a central tower for aviation navigation.

To the SE:

-The **mill and chimney of Estate Morningstar** are inland in the Salt River Valley. Almost in line with them near the top of a hill to the southeast is the **chimney of Estate Rattan**.

-**Hess Oil Refinery** on the south shore. The **chimney of Estate Diamond** is lined up with a gap between rows of the Hess storage tanks.

0.3 Fork at plot 81. Go left.

0.45 End of loop at mile 1.8(13.2) of Route 78

<0.4>

Scenic Drive East, Route 78 (continued)

E-W (W-E)

2.6 (12.4) Fork. To continue Route 78 take the left fork and veer S. A dirt road runs straight or NW here, and when bush is cut, it gives good **views of both coasts from a ridge of 720 feet elevation**. The road is about 200 yards long and pleasant to walk.

<1.35>

Canaan Ridge Road, Route 73:

3.95 (11.05) Junction with paved Route 73 [p. 132, mile 1.3(3.4)]. To continue on the Scenic Drive, make a right or N turn here, and drive Route 73 for 0.15 mile.

<0.15>

Scenic Drive West, Route 78

4.1 (10.9) Junction of Route 73 at mile 1.15(3.55) with the start of Scenic Drive West, a paved road that turns off to the left or W. Turn left here to continue the Scenic Drive, Route 78. This next section of the scenic drive runs on Eagle Ridge, the highest mountain on the island.

Side Trip on Scenic Drive, Route 78

Scenic Drive West, Route 78

E-W (W-E)

<0.25>

4.35 (10.65) Dirt road to right or N, Vallee View Road. This road goes 1.25 miles to North Shore Road, Route 80.

Vallee View Road (S-N):

0.0 Junction with Scenic Drive West at mile 4.35(10.65)

0.45 Dirt road to left or W, residential with **view**

E-W:

0.0 Junction with Vallee View Road at mile 0.45

0.15 At a fork, **View of Maroon Ridge**, ending in Ham Bluff with lighthouse in distance, and Cane Bay below. To the E one can see the **mill of Rust-op-twist** and Baron Bluff, the peaked hill on the shore.

0.2 Road end

0.55 Start of pavement

0.65 Junction. Turn left or W. The **view to W** from this point is of Maroon Ridge, Ham Bluff Lighthouse, and Cane Bay below.

1.25 End at Route 80 [p. 110, mile 9.7(9.8)]

<0.25>

4.6 (10.4) Fork. Take the left fork to continue Route 78.

<0.3>

4.9 (10.1) End of pavement

<0.7>

5.6 (9.4) Dirt road to left or S. Locked gate at 0.2 mile.

<0.05>

5.65 (9.35) On the left or S side of the road is a clump of six large trees. These are native **West Indian Locust Trees**. This tree is most commonly found in higher elevtion moist forests, but its seed pod is common among drift seeds on shores. If any pods are on the ground, you will see how durable they are, enough for a wash down a mountain river and then a float for several weeks in the ocean. The leaflets are emerald green with a curious paired arrangement. These trunks are covered with lichens, many of them orange. A lichen is an alga and fungus growing together. The alga produces food for the fungus, and the fungus provides moisture and shelter for the alga. The middle trunk has several rings of **Yellow-bellied Sapsucker holes** about 20 feet off the ground. [See p. 103, mile 4.6(14.9).] Look for resin oozing from wounds in the trunks. This is South American copal, used as incense by the Mayans and in varnish. Sap deposits of a close relative that lived 20 million years ago turned into the New World fossil amber that is so highly prized for jewelry today.

<0.85>

6.5 (8.5) **Viewpoint to south shore** at 700 feet elevation. Bush sometimes obscures the view. Sweeping left to right (E-W):

Scenic Drive West, Route 78
E-W (W-E)

-Radio and transmitter towers on Blue Mountain
-Industrial complex along the south shore
-Krause Lagoon mangrove with the dredge spoil Ruth Island to south
-Tall, circular **chimney of Bethlehem Old Works** sugar factory, inland
-Red-roofed buildings of University of the Virgin Islands

<0.4>

6.9 (8.1) Turnout on right or N. Walking up to the top of the excavated area, one can view Maroon Ridge ending in Ham Bluff with the lighthouse on top.

<0.3>

7.2 (7.8) Junction with North Shore Tour, Route 69 [p. 114, mile 11.7(7.8)]. To continue the Scenic Drive, stay on dirt and drive straight or W to cross Route 69.
-River Road, Route 69 [p. 133, mile 0.75(3.65)], goes 0.8 miles to right or N to the north shore. To the left or S, it connects in 2.35 miles with Midland Road, Route 72, and Centerline in 3.6 miles.

SIDE TRIP ON CANAAN RIDGE ROAD, ROUTE 73

(Also Mon Bijou Road, LaVallee Road)

This road runs from the village of LaVallee on the North Shore Road, Route 80, up Canaan Ridge through moist forest resulting from heavy runoff from Blue Mountain. At 3.3 miles south it intersects Midland Road, Route 72, and at 3.95 miles ends at Route 707. From here this log continues south on Route 663, intersecting Centerline at 4.7 miles and the Four-lane Highway, Route 66, at 5.4 miles.

N-S (S-N)

0.0 (4.7) Junction with the North Shore Tour on Route 80 at the village of LaVallee [p. 109, mile 8.65(10.85)]. A green mileage sign "4" is near the junction. The Sea Shadows **SCUBA and Dive Shop** is on the SW corner.

<0.3>

0.3 (4.4) **Mill and ruins of Estate Belvedere** incorporated into private residence on left or E. At this mileage there is a lovely wall of coral block showing cross sections of brain coral cut to make the stones. When the coral was alive, the spaces were filled with tissue of living animals, each track filled by an individual called a polyp, a tiny animal similar to a sea anemone, but which has the ability to deposit calcium carbonate rock around its body and thereby make a coral rock. Look at the bottom center of the fan-like pattern to see how the brain coral colony starts from a single polyp, which branches into two polyps. These two

N-S (S-N)

in turn divide into two each, and each of those four into two each, and so on. The original polyps keep growing lengthwise, dividing, and depositing skeleton, and the whole process results in a coral colony with this intricate pattern. Each species of coral has a different pattern of framework construction.

<0.85>
1.15 (3.55) Junction with Scenic Drive West, Route 78, [p. 129, mile 4.1(10.9)].

<0.15>
1.3 (3.4) Junction with Scenic Drive East, Route 78 [p. 129, mile 3.95(11.05)].

<0.2>
1.5- (2.15-
2.55 3.2) **Wet forest area** due to runoff from Blue Mountain

<0.05>
2.6 (2.1) **Chimney and ruins of Estate Libanon** to left or E

<0.1>
2.7 (2.0) Junction with paved road to left or E, dirt to right or W. The paved road goes 1.65 miles, passing an area with small farms, to join Route 80. [See "Glynn Farms Road", p. 102, mile 1.65(0.0).]
-The dirt road to W goes to private **Estate Little Fountain**.

Road to Estate Little Fountain (E-W):

0.0 Junction with Canaan Ridge Road at 2.7(2.0)

0.1 Two large **Gregre Trees** are on the south side of the road on either side of a concrete building. The numerous rings of small holes on the trunks were made by **Yellow-bellied Sapsuckers**. [See p. 103, mile 4.6(14.9)].

0.3 **Mill of Estate Little Fountain** on the left or S. This mill has retained much of its plaster coating, but is split in half, the southern side a rubble heap. The road ends at a private house at 0.4 mile.

<0.3>
3.0 (1.7) **Ruins and mill of Estate MonBijou** to right or W incorporated into private residence

<0.3>
3.3 (1.4) Junction with Midland Road, Route 72 [p. 88, mile 8.75(0.9)]. A left or E turn goes 0.9 mile to Route 75. A right or W turn goes 8.75 miles, much of it through **moist forest**, to the West End Road, Route 63, at the western coast.

<0.65>
3.95 (0.75) End of Route 73. Turn left or E onto Route 707. **Ruins and mill of Estate Fredensborg** are hidden in the brush on the NW corner of this intersection. Straight or S across the intersection is a rough dirt road (four-wheel drive) to inhabited ruins of Estate King (also called Upper

Canaan Ridge Road, Route 73
N-S (S-N)

Bethlehem) and Centerline [p. 29, mile 7.5(7.9)]. Right or W goes 0.4 mile to Fredensborg Pond and on to ruins of Bethlehem Old Works sugar factory. [See map on p. 31.]

<0.05>
4.0 (0.7) Turn right or S onto Route 663. Route 707 continues straight or E for 0.3 mile to join Route 75 [p. 101, mile 5.55(0.6)].

<0.3>
4.3 (0.4) A **watermill** of Estate Fredensborg is on the left or E.

<0.1>
4.4 (0.3) Veer right or W. On the right or N are **ruins and a large mill of Estate Slob** (also Body Slob). The ruins are chiefly of the village, arranged in the "old-fashioned" (early 1700s) manner with separate little houses for each of the slave families. Later, estates economized by building the longhouses. Examples of the longhouses are at villages of Estate Bethlehem New Works (nearby with moderately long apartments: p. 55, mile 0.5(5.25)) and Estate Hogensborg with really long buildings [p. 37, mile 12.2(3.2)].

<0.3>
4.7 (0.0) Junction with Centerline, Route 70 [p. 28, mile 7.4(8.0)]. Sign for northbound vehicles on NW corner near the Kingshill Lutheran Church says "To 707". [A log for Route 663 to the south of this point, 0.8 mile to the Four-lane Highway, is given in the Centerline description.]

SIDE TRIP ON RIVER ROAD, ROUTE 69

(Also Fountain Valley Road)

This road goes 4.4 miles between the North Shore at the Davis Beach Junction to Centerline, Route 70. It provides access to the most popular western section of the Scenic Drive and to good birdwatching at the freshwater ponds of the Carambola Golf Course.

N-S (S-N)
0.0 (4.4) Junction at terminus of Route 80 and Route 69 [p. 113, mile 10.95(8.55)]. The Davis Beach road goes right or W from here to the hotel complex of Carambola Beach Resort.

<0.75>
0.75 (3.65) Junction with dirt Scenic Drive West, Route 78 [p. 114, mile 11.7(7.8) for the right or western section, and p. 131, mile 7.2(7.8) for the left or eastern road.] The knoll on the NE corner of this junction provides a **nice view of the northern Virgin Islands** on a clear day. The knoll on the W has a dirt road (no vehicles allowed), which climbs to **scenic views of both shores** from Bennykenny Hill (713 feet) and then descends steeply (20% grade) through moist forest to Davis Beach.

N-S (S-N)

<0.15>

0.9 (3.5) **Viewpoint** to right or W down into Carambola Golf Course in Fountain Valley. The **mill of Estate Big Fountain** is in the valley near the clubhouse. The **mill of Bodkin** stands out on the ridgeline above.

<0.4>

1.3 (3.1) Access to **mill of Estate Parasol** to left or E. The access point is across the road from an area of the golf course where grassy fields change to wooded forest on a small ridge crest. There is a wire gate in the fenceline. Walk left or N up the dirt powerline road just inside the gate to the mill (about 150 yards). Many ruins of the old estate are overgrown in the forest south of the mill.

<0.3>

1.6 (2.8) Paved road to right or W to Clubhouse of Carambola Golf Course. This golf course has many horticultural plantings.

E-W:

- 0.0 Junction with Route 69 at mile 1.6(2.8). **Honduran Mahogany Trees** line the south side of the road for 0.15 mile up to a golf cart crossing.
- 0.45 **White-cedar Trees** line the right or N side of the road for 0.05 mile starting just after some speed bumps.
- 0.65 Veer right or N through **ruins of Estate Big Fountain** with a line of **Gumbo-limbo Trees** on the right or E. You are welcome to explore the picturesque ruins and mill. You can shoot some wonderful photographs of **Strangler Figs** doing just that to the ageing structures.
- 0.75 Clubhouse and **restaurant**. A map of the course is on a wall of the clubhouse. Ask here at management desk for **permission to birdwatch** on golf course ponds. [Access is at mile 2.45(1.95) below.] **Ruins of mill of Estate Big Fountain** are on a hill to W.

<0.85>

2.45 (1.95) Dirt road to right or N is main access to ponds of golf course and **good birdwatching area**. To birdwatch ask permission at clubhouse--see mile 1.6(2.8).

S-N:

- 0.0 Junction with Route 69 at mile 2.45(1.95). At about 25 yards from Route 69 one comes to a fork. Go left. The road to right leads into an area with several **groves of exotic trees**. This was the nursery area for the golf course. In a line along Route 69 are **Silver Tabebuia Trees.**
- 0.05 Fork. Go left.
- 0.15 End of road at **freshwater ponds** at southern end of the golf course.

<0.05>

N-S	(S-N)	
2.5	(1.9)	Bridge crossing of Jealousy Gut. Very large **Raintrees** shade this area. The curtains of reddish roots are from *Cissus* vines, which cover much of the canopy.
<0.1>		
2.6	(1.8)	**Ruins of the village of Estate River** on both sides of the road
<0.1>		
2.7-3.0	(1.4-1.7)	Large **West Indian Mahogany Trees** line the left or E side of the road, starting near a bridge about 200 feet S of the green mileage sign "2.5".
<0.15>		
3.15	(1.25)	Junction with Route 72 (Midland Road) [p. 86, mile 6.25(3.4)]. To continue Route 69 go straight or S. Turning left or E takes you 3.4 miles to Route 75.
<0.35>		
3.5	(0.9)	Turn left or S to continue Route 69. Continuing straight or W down Mahogany Road, Route 76, one goes 5.95 miles to junction with West End Road, Route 63, on the west coast.
<0.3>		
3.8	(0.6)	Paved road to left or W, Route 705
<0.15>		
3.95	(0.45)	Dirt road to right or W. **Mill of Estate Adventure** can be seen on hill to S ahead, across Centerline Road.
<0.45>		
4.4	(0.0)	Junction with Centerline, Route 70 [p. 33, mile 9.8(5.6)].

Fig. 44. Tree Lizards engaging in a favorite activity.

Fig. 45. Spear fishermen in Bluegut Bay divide their catch.

EAST END TOUR

Including East End Road, Route 82, Southshore Road, Route 60, and Southside Road, Route 62

This tour runs along the coastline of the eastern half of the island which is quite different from the western half because of its dryness and its less dense population. The eastern part of the island receives only about 20 inches of rain each year, while the western end recives about 40 inches. Hence, the typical vegetation of the eastern end is scrub acacia and cactus in marked contrast to the evergreen forests of the west. Vistas along the north coast are nearly all punctuated by the two islets, Green Cay and Buck Island. East Point is noted as "the easternmost land in the U.S.A." Most of the eastern tip is a territorial park, with about a mile of pristine, white sand beach, accessible only by hiking after driving over dirt roads for about 1.5 miles. This tour begins at Christiansted and follows the northern coastline from W-E for 11.7 miles on East End Road, Route 82. It then follows the southern coastline from E-W along Southshore Road, Route 60, and Southside Road, Route 62 for 14.3 miles, ending near the middle of St. Croix, about two miles W of Christiansted.

EAST END ROAD, ROUTE 82

(also Northeast Shore Drive)

W-E (E-W)

0.0 (11.7) Christiansted Harbor parking lot in front of the Old Customs House. Seaward, about 30 yards further E is **Fort Christiansvaern**, and Company Street is on the inland side. This is the beginning of the "Centerline Tour" [p. 17, mile 0.0(15.4)].

<0.15>

0.15 (11.55) Left or seaward is a gas station on the NW corner of a one-way road that goes past the **fish market of Christiansted** at Gallows Bay. The fishermen congregate with their boats, scales, and colorful catches in an area of trees and sand, right along the shore. A small wooden pier in the market area is at 0.1 mile from the junction with Route 82. The most active times are Wednesday and Saturday mornings starting at about 9am.

<0.35>

0.5 (11.2) Paved entry to dirt road on right or S to Recovery Hill and **view of the middle section of the island**. A gas station is on the NW corner of the intersection.

Road to Recovery Hill (N-S):

0.0 Junction with Route 82 at mile 0.5(11.2). At a three-way fork, about 50 yards from junction, take the middle dirt road, veering slightly left and uphill.

0.4 Junction with another dirt road. Turn left.

1.3 Three-way fork. Take the right-most branch.

1.55 View of the north shore

1.6 **Good view of Christiansted Harbor**. At 30 yards further is a four-way fork with a **good view of the south shore** at the junction.

Including East End Road, Route 82,
Southshore Road, Route 60, and Southside Road, Route 62

East End Road, Route 82

Recovery Hill (continued):

View of Christiansted Harbor, moving from W-E (left to right):

-The highest part of St. Croix, Eagle Ridge, with its two tallest peaks, Blue Mountain (with antennas, 1096 feet) and Mount Eagle (1165 feet)

-The peninsula that extends out to sea in the west is Estate Judith's Fancy. West of the peninsula on the skyline is Saltriver Mountain with two peaks, Windsor Peak (885 feet) and Clairmont Peak with a house (866 feet).

-Proceeding east from Judith's Fancy, one can see **Christiansted Harbor**, formed by the barrier coral reef called **Long Reef**.

-The power plant and large green water tanks are **facilities of the Water and Power Authority** (called by its acronym WAPA) which provides all electrical power for St. Croix. It is built on the ruins of Estate Richmond, which had the first communal sugar factory on the island, begun and terminated in 1877. A deep green, dredged channel runs from the mouth of Christiansted Harbor to the WAPA dock.

-In line with the radio station antenna is submerged **Scotch Bank** offshore, a dark brownish green surrounded and riddled by the lighter turquoise blue of sand bottom.

-**Altona Lagoon** is the large mangrove which lies between the radio tower and the Buccaneer Hotel, a large pink structure on a hill to the east.

-In the distance are the islets Green Cay and Buck Island. Pull Point is in line with Buck Island viewed from Recovery Hill. It has several large stone buildings, and it is joined to the main part of St. Croix by an isthmus.

-The sandy island beyond Pull Point is Candle Reef.

-Just east of Candle Reef one can see the westernmost end of Tague Bay Reef.

-Two peaks of the Goat Hills are visible on a clear day. The furthest is the highest, Goat Peak, 660 feet. Inland and closer is the Seven Peaks Range. The closest is the highest, Peak Seven, 860 feet.

-When it is clear you can easily see the northern Virgin Islands.

View of the south shore, left to right (E-W):

-Halfpenny Bay

-Manchenil Bay, with the **mill of Estate Diamond Keturah** on an inland rise at the Bay's midpoint.

Recovery Hill (continued):
-Vagthus Point (Danish) or Watchho (English), the elevated peninsula forming the western boundary of Manchineel Bay
-Hess Channel, often with large tankers at dock
-Ruth Island, created by dumping dredge spoil on a coral reef when the Alucroix Channel was made
-Krause Lagoon, inland from Ruth Island. The Alucroix Channel runs on the eastern side of Krause Lagoon.
-Hess Oil chimneys (black and silver striped) and domed tanks of aluminum refinery (now closed)
-On a clear day: Long Point; the Bight of Long Point; Westend Saltpond; and Sandy Point, the western tip of the island.

-The right-most fork here goes to the top of Jacobsberg Ridge, Recovery Hill (845 feet), which has a splendid, nearly 360° view of Christiansted and the middle of the island. However, you may need permission to enter because of the airport navigational equipment that occupies the hilltop. Hermon Hill Road [p. 22, mile 1.3] leads to a similar view but from a more accessible site on Jacobsberg Ridge.

<0.15>
Continuing on Route 82:
W-E (E-W)
0.65 (11.05) Junction on right or S with Route 85, Springgut Road. This road goes 2.4 miles to Southside Road, Route 62 [p. 170, mile 10.4(3.9)]. A stadium is on the SW corner.

<0.1>
0.75 (10.95) Paved road to left or N to Altona Lagoon and Fort Louise Augusta area.

Road to Altona Lagoon (S-N):
0.0 Junction with Route 82 at mile 0.75(10.95)
0.05 **Mill of Estate Mt. Welcome** is on a hill to left or E.
0.4 Junction. Turn right at first view of Christiansted Harbor. The bridge you cross shortly after the turn is the inlet to **Altona Lagoon. Fishermen** are often here using cast nets for bait fishing.
0.5 Path to southeast along border of inlet. The lagoon and inlet are principle **nursery areas for juvenile fishes.** Some common ones that you may see (using polarized sunglasses to look into the water): juvenile snappers; mullets with dark edged tail fins; halfbeaks, swimming near the surface, with their long protruding lower jaws; needlefish (locally called "gar"), which also swim right at the surface; and

Road to Altona Lagoon, S-N (continued):
damselfish, which guard territories on the rubble in the bottom of the inlet. Juvenile **Great Barracudas**, one foot and smaller, are common. The **Upside-down Jellyfish** is abundant on the bottom. It usually pulsates its body, revealing its presence. Many kinds of **birds** are abundant here.

Fig. 46. Red Mangrove Trees line the inlet of Altona Lagoon.

-If you are stalwart, Altona Lagoon makes an interesting and different kind of snorkel from those over coral reefs. (You will need to be well covered with long-sleeved shirts and long pants because the Upside-down Jellyfish have tentacles trailing everywhere. No matter how covered you are, you will probably receive a few stings.) Peering beneath the Red Mangrove roots you will see many juvenile fishes, especially **Schoolmasters** with their yellow body, dark vertical bars, and stripe through the eye, and **Great Barracudas**. The roots in the water are covered with marine life including orange and yellow sponges, seasquirts, stalked anemones, and oysters.

0.55 Path to border of inlet of Altona Lagoon

0.75 Building on left or W has shade for **picnicking and public restrooms**.

Road to Altona Lagoon, S-N (continued):

0.8 Dirt road to right or E to view (W-E):

- 0.0 Junction at mile 0.8 of road to Altona Lagoon
- 0.05 Border of Altona Lagoon on SE
- 0.1 Parking area for Fort Louise Augusta Beach (also called Camphouse Cove). The shore is lined with **Seagrape Trees** which have edible purple fruits in summer. The rocky shore that borders the east side of the lagoon has some corals and small fish.
- 0.2 Hilltop **view of Altona Lagoon and surrounding salt flats**. Use the parking area for the beach and walk up the hill. From this hill one can see the following points, swinging from W-E.
  -Judith's Fancy, the headland to the west.
  -Mount Eagle, the tall mountain inland (1165 feet) is the highest point on St. Croix. Further inland is Blue Mountain (1096 feet) with numerous antennae.
  -Long Reef, the barrier that forms Christiansted Harbor
  -WAPA (Water and Power Authority) with its large green water tank
  -Radio station WSTX on ruins of Fort Louise Augusta
  -Beauregard Beach and Bay border the northern shore of Altona Lagoon. The rocky promintory with buildings of the Buccaneer Hotel that juts out into Beauregard Bay is Shoy's Washer. The little cove east of the Washer is Martel Bay with Shoy Point forming its eastern edge.

0.9 Road end at **ruins of Fort Louise Augusta** and **radio station WSTX**. The fort was built in the early 1600s. These remnants represent the oldest structure on St. Croix. Park and walk seaward for a **good view of Christiansted Harbor**. Visible here are Protestant Cay with Hotel on the Cay, Fort Christiansvaern, boats anchored in Gallows Bay, and the mill of Estate Mount Welcome on the hill just east of Gallows Bay. Recovery Hill with a short tower is visible south of Mt. Welcome Hill.
-Walk east along the Fort to view Camphouse Cove just below, Beauregard Beach beyond to the east, and Martel Bay just beyond that. Shoy Point is the eastern edge of Martel Bay. Green Cay juts out into view east beyond Shoy Point, and Buck Island is in the distance further east and out to sea.

Including East End Road, Route 82, Southshore Road, Route 60, and Southside Road, Route 62

East End Road, Route 82

Road to Altona Lagoon, S-N (continued):

-Walk up the hill to the radio tower. Here one can see all of Altona Lagoon. Looking north out to sea, the dark brownish-green area is Scotch Bank.

<0.6>

1.35 (10.35) Junction with Route 62 to right or S. A gas station is on the left or N side at the junction. This road joins the East End Tour, Route 62 section, 1.9 miles away. There is a **good viewpoint** at the top of Maria Hill at 1.2 miles. This section of Route 62 is described as "Lowry Hill Road", [p. 169, mile 1.9(0.0)].

<0.2> <0.3>

1.55 --- **Mill of Estate Butzberg** on hill to right or S (private)

--- (10.05)

<0.55><0.45>

2.1 (9.6) Paved road to left or N leads to **Buccaneer Hotel** and Estates Shoy and Anna's Hope (which has Darby Hill). The Buccaneer Hotel has access to several nice **sandy beaches** and **good birdwatching areas along the shores of Altona Lagoon and freshwater ponds** on the golf course. **Viewpoints** on the residential roads overlook Christiansted, Altona Lagoon, and Southgate Saltpond.

Road to Darby Hill and Buccaneer Hotel (S-N):

0.0 Junction with Route 82 at mile 2.1(9.6)

0.1 Guardhouse for Buccaneer Hotel.

(a) **Darby Hill**: A right turn onto the road just outside the guard house leads to viewpoints.

W-E:

0.0 Buccaneer guard house. Turn right before entering the Buccaneer grounds.

0.2 Fork. Take the right fork onto dirt.

0.5 Fork. Go right. A left turn takes you back to the fork at 0.2.

0.65 Junction with road to Darby Hill (Buccaneer Bay Estates). Turn right or S for views.

-Just opposite this junction, a dirt path goes seaward to the quiet, often deserted **white sand beach of Punnett Bay** (often called Shoy's Beach). Despite the sign, the path is not dangerous. Park along the road or on the corner of a vacant lot across the road. When conditions are right there is good **body surfing** here. **Leatherback Sea Turtles** nest on Shoy's Beach. The evidence she leaves is a huge track that looks like someone drove a tractor up out of the water and back in again. [See p. 59, mile 2.95.]

Road to Darby Hill, W-E (continued):

-Another beach access is at the end of the road, 0.25 mile further. A path to the left of the road end runs between two houses.

Continuing on Buccaneer Bay Estates road:

0.7 Three-way fork. Take the middle road uphill.

0.75 Fork. Go right uphill.

0.9 **View of Southgate Saltpond** to the east.

1.1 Fork. Go left downhill.

1.2 **A fine view of Altona Lagoon and Christiansted Harbor** You may continue this road to loop back to where you entered.

(b) **Buccaneer Hotel**. To enter the Hotel grounds, stop at the guard house and state the purpose of your visit. The beach facilities fee is $3 per person.

S-N:

0.0 Guard house.

-About 25 yards beyond the guard house just in front of the vine-covered **chimney ruin** is a **Baobab Tree**, also called "Dead Rat Tree" because of its large, pendulous fruits, which form towards the end of summer.

0.2 Fork. Go left at this fork for parking for the beaches and Altona Lagoon. Right leads to the main desk of the hotel, and most of the **ruins of Estate Shoy** (also Shoy's) including a **mill**. The desk provides **maps of the grounds**. Some of the plants around this area of the hotel are labeled. There are plans to organize the **horticultural collection**, labeling more of the plants and providing a guide map for them.

0.25 Parking for beaches and hotel grounds. The parking lot at the restaurant just uphill provides a **good view over Altona Lagoon**.

(1) **The Martel Bay Area**:

The hotel has named the beach in view of the parking lot "Mermaid Beach" and the little bay "Cutlass Cove." The traditional name is Martel Bay--"martel" is French for "hammer". The beach house **rents and sells equipment and dive-snorkel tours**. The rocky point that forms the west end of the bay is called "Shoy's Washer", perhaps because of pounding surf on the point in winter months. Shoy Point is the headland on the eastern border of Martel Bay. The **view from the point** is very scenic. To get there, walk uphill along the shoreward side of the Doubloon Rooms, which are east of the

Buccaneer Hotel, S-N (continued):

Beach Shack. Take the golf cart path towards hole 3 and Shoy Point. Paths through tall grass lead off towards the Point from the golf cart path. Further east of Shoy Point is the **isolated, sandy beach** named "Whistle Beach" by the hotel. One can walk along the eastern shore of Whistle Beach out to the rocky western point of Punnett Bay.

Fig. 47. Beauregard Beach, with its typical crowd.

(2) **Beauregard Beach Area** (called "Grotto Beach" by the hotel) and **Birdwatching at Altona Lagoon Salt Flats**:

West beyond the Beach Rooms is the **half-mile, sandy stretch of Beauregard Beach**. One can walk along the beach and over the bedrock on the west to Camphouse Cove below the radio station of WSTX. To reach the **salt flats of Altona Lagoon**, walk to the western end of Beauregard Beach. Here there is a broken fence line along a hill, with paths leading through the bush to the salt flats, about 100 yards south of the beach.

(3) **Birdwatching along border of Altona Lagoon and the Buccaneer Golf Course**:

Buccaneer Hotel, S-N (continued):

Shorebirds often feed on the lawns of the golf course. At the western end of the Beach Rooms is the Grotto Restaurant. A dirt road on the west side of the buildings runs along an arm of the lagoon. Near hole 7, the water turns clear and one can usually see the **Upside-down Jellyfish** pulsating on the bottom. **Mullets** are numerous. They are elongate fish with a dark margin on their back and tail fins, and they usually swim in schools slightly below the surface. This part of the inlet is lined with **Red Mangrove Trees** and **Black Mangrove Trees**. The dirt road continues along the lagoon margin, a good place to look for wintering warblers. Follow signs to hole 14 to access some arms of the Lagoon.

(4) **Birdwatching at freshwater ponds**:
Several ponds are located close to the guard house at the entrance gate. The freshwater pond near the entrance at Route 82 is often a good one for birdwatching. Go around the Golf Shop, staying on the golf course side of the fence. The largest freshwater pond on the course is at the extreme southeast corner of this part of the golf course. It is often a good bird area because of its isolation.

<0.95>

East End Road, Route 82 (continued)

W-E (E-W)

3.05 (8.65) Paved road to left or N to Tamarind Reef Beach Club, Green Cay Marina, and hike to **Southgate Saltpond and Southgate Beach**. **Green Cay**, which is just offshore here, is a **National Wildlife Refuge** for the endangered **St. Croix Ground Lizard**. The lizard is abundant on the beaches of the Cay, and occurs sparsely in other habitats on the island. Green Cay has a dense forest on the southern end that provides a rookery for pelicans and herons.

Tamarind Reef Road (S-N):

0.0 Junction with Route 82 at mile 3.05(8.65)

0.55 Marina Road to right or E at a gate in fenceline. Green Cay Marina is open to the public, but to use the road to Southgate Saltpond and Southgate Beach, one must pay the beach-use fee to the Tamarind Beach Club [see mile 0.9, below]. You must park at the Marina and walk 0.5 mile to the pond and beach.

EAST END TOUR
Including East End Road, Route 82, Southshore Road, Route 60, and Southside Road, Route 62
East End Road, Route 82

W-E (E-W)

Tamarind Reef Road, S-N (continued):

Marina Road to Southgate Saltpond and Beach (W-E):

0.0 Green Cay Marina gate at mile 0.55 of Tamarind Reef Road. Turn right just inside the gate. This road is chained off to vehicles.

0.1 Fork. Go left uphill. This road runs along the west border of **Southgate Saltpond** with good views of birds from under cover of the mangrove trees.

0.4 Turn right along the pleasant **white sand beach of Southgate** with shade of **Seagrape Trees.**

0.5 Road ends near some concrete ruins. The saltpond access is by overgrown paths along the pond edge through brush and mangrove forest.

Tamarind Reef Road, S-N (continued):

0.9 Road ends at Tamarind Reef Hotel and Beach Club. The Hotel charges $3 per person for use of its beach and pool facilities. The fee includes use of the Green Cay Marina roads. The beach near the hotel includes a sheltered salt water cove on the west side of the main jetty of the marina.

<0.2>

3.25 (8.45) **Greathouse of Estate Southgate** to left or N (private)

<0.1>

3.35 (8.35) Junction with Route 60 a paved road to right or S. [See "Tipperary Road", p. 175, mile 0.0(1.5).] At 1.5 miles it rejoins the East End Tour on the south shore. Route 60 provides access to some **scenic viewpoints** of this part of St. Croix.

-To continue the East End Tour, turn left or N here.

-At 15 yards from the junction, on the right or E is the entrance to Southgate Farm Condominium. The **ruins and mills of Estate Southgate** are near this entrance.

<0.25>

3.6 (8.1) Small building on inland side. The trees in front with deep green leaves are **Common Lignumvitae.**

<0.2>

3.8 (7.9) Dirt road seaward goes to **white sand beach** bordering Southgate Saltpond. The beach has pleasant shade from **Seagrape Trees.** One can walk into the northeastern corner of the saltpond here for **excellent birdwatching.** Birds seen frequently include White-cheeked Pintail, White-crowned Pigeon, Brown Pelican, herons, egrets, warblers, plovers, gulls, and terns.

Southgate Saltpond (S-N):

0.0 Junction with Route 82 at mile 3.8(7.9)

Including East End Road, Route 82, Southshore Road, Route 60, and Southside Road, Route 62

East End Road, Route 82

W-E (E-W)

Southgate Saltpond, S-N (continued):

0.15 Dirt road to left along the southern margin of the mangroves, runs 0.2 miles. This road is quite soft when wet, and one should park and walk.
-About 50 yards further is a junction.
-Drive left or W for **multiple beach access and parking areas** ending in 0.15 mile. At 0.1 mile there is a path inland towards the pond, about 15 yards E of (before) an access N to beach. Walk in through a mangrove clearing to the NE corner of the saltpond.
-Drive straight or right for **beach access**. In the wave-wash zone along this beach is a shelf made of consolidated fossil corals and other reef organisms. The seaward face has many little nooks and crannies for sealife, and is a good nearshore **snorkel** site. The seagrass bed offshore is principally composed of thin, round-bladed **Manatee Grass**. The broad-bladed grass is **Turtle Grass** (the main component of most seagrass beds). If you try to roll some of the thin blades between your fingers and find they are flat (with edges) and not round, then you have picked some of the less-common **Shoal Grass**. None of these three species is really a grass, but they are all flowering plants, not seaweed (algae). They have rather unusual flowers, but mostly spread by extending their root systems.

<0.3>

4.1 (7.6) Paved road seaward to Chenay Bay

0.0 Junction with Route 82 at mile 4.1(7.6). The **chimney, windmill, and factory ruins of Estate Greencay** are on the hill above.

0.2 Parking lot for Chenay Bay Colony, a hotel, is open when the restaurant is open. A **watermill** is on the right or E of the parking area. Park and walk to the **white sand beach** of Chenay Bay. There is no fee.

<0.2>

4.3 (7.4) Paved road to left or N, Miss Bea Road. This road provides access to a **white sand beach, Prune Bay**.

Miss Bea Road (S-N):

0.0 Junction with Route 82 at mile 4.3(7.4). About 15 yards from the junction is a dirt road to left or W that passes the **chimney, factory ruins, and mill of Estate Greencay** (private) at 0.1 mile distance.

0.2 Road to right. Go straight.

0.4 Paved road veers left or W. Go straight onto dirt road.

0.45 Fork. Go right.

W-E (E-W)

Miss Bea Road, S-N (continued):

0.5 Isthmus of Pull Point. Access to right for the beach of Prune Bay. A coconut grove provides shade.

0.55 Turnaround area for Pull Point. The stone buildings on the end of the Point are private.

-The path to left or W leads to a protected cove (Heathen Bay) where boats were traditionally hauled for work--hence the name of the point. The beach is a mixture of sand and smooth cobble. The bedrock forms some **tidepools** in the western part of the cove. One can walk southwest from here towards Chenay Bay beach, but in the last 100 yards you will have to get wet at least to the waist to arrive at the beach.

-To the right is the **lovely, usually deserted beach of Prune Bay** with a shady coconut grove nearby. From this point one can walk east as far as the rocky point you can see at the end of the beach. One needs hiking shoes to scramble over the rocks on that point to the beaches of Coakley Bay on the other side. A fringing coral reef is just NW of the point, nice for **snorkeling**.

<0.2>

4.5 (7.2) Paved road to right or S. This road is lined with **Lignumvitae Trees** on the left or E and **White Manjack** on the W. The **old Danish Easthill Schoolhouse** on the hill above is now a private home.

<0.8>

5.3 (6.4) Dirt road to left or N goes up onto Wismenog Point to **ruins and mill of Estate Coakley Bay**. This road is also access to the **shaded, white sand beaches** of Coakley Bay, and **Coakley Bay Saltpond, a birdwatching area**.

Road to Coakley Bay Saltpond (E-W):

0.0 Junction with Route 82 at mile 5.3(6.4)

0.05 The main road veers left. A dirt road climbs straight uphill to the **mill and greathouse ruins of Estate Coakley Bay**. The Danish had a small battery here in the late 1700s and early 1800s. It was one of seven shore batteries, each with a few cannon and men. Others were at Grieg Hill on Salt River, Protestant Cay, Fort Louise Augusta near Christiansted, Long Point, Prosperity near Frederiksted, and Sprat Hall.

-You can walk down to Carden Bay, the small cove just east of the mill. On the east end of the greathouse just south of a lime tree, turn right or E and walk on a path through the grass for about 25

Road to Coakley Bay Saltpond, E-W (continued):

yards. Then as another part of the ruins comes into view (about 15 yards away), turn left or N and walk down to the cove.

0.1 Start of a line of **ruins of Estate Coakley Bay**, both sides of the road. The ruins include a village, factory, and circular-walled animal mill.

0.2 Beach access, shaded by Manchineel Trees.

-An **interesting snorkeling area**, a reef with many sponges, sea fans, sea whips, and other soft corals, trends NW off the point with the mill on top. It is only 10 feet deep, and about 100 yards long.

-From this point one can walk W as far as Pull Point and E as far as Tague Bay. Wear shoes you can wade in. One of the most difficult places in the walk to the east is right at this beach. To stay dry here, one must crash through some brush. Wading around the brush in the water is easy. The next cove (Carden Bay) to the E has lots of **Touch-me-not** [see p. 179], so heed the warning and do not hold on to brush indiscriminately.

0.25 **Large Strangler Fig Tree** to the left or S or the road, with multiple trunks and prop roots. [See p. 75, mile 4.75(2.2) for more about Strangler Figs. This is the Shortleaf Fig.] This tree is useful in finding the E entrance to Coakley Bay Saltpond, a **birdwatching area**. Count three beach-side access roads past this fig, and nearly opposite the third is a small cleared area on the inland side, mile 0.35 below.

0.35 Beach access to north, path to saltpond to south. A path leads SE through the manchineel forest to the eastern end of the saltpond.

0.45 Road ends at a beach access point. The sandy islet offshore is Candle Reef (also Dry Ledge). From this clearing, walk W along the shore about 150 yards until you come to a clearing of grass and vines, about 25 by 50 yards. Pull Point will have just come into view, and to continue down the shoreline to Prune Bay you will have to scramble over the rocks of this point. At about the middle of the clearing, a path leads back through the manchineel forest to the western end of Coakley Bay Saltpond. When the pond is fairly dry, one can walk along the southern shore through the Black Mangrove Trees all the way to the eastern entrance at 0.35 mile above. The northern edge of the pond is wet and densely lined with Red Mangrove Trees.

EAST END TOUR

Including East End Road, Route 82, Southshore Road, Route 60, and Southside Road, Route 62

East End Road, Route 82

W-E (E-W)

<0.3>

5.6 (6.1) Coakley Bay Condominium to right or S

<0.05>

5.65 (6.05) Dirt road access to **sandy beach of Hodge Bay**, with Seagrape Tree shade at less than 0.05 mile. One can walk from here west to Pull Point and east to Tague Bay wearing hiking shoes and wading a minor amount.

<0.1>

5.75 (5.95) Parking area seaward near a hog-wire fence. Walk down wooden stairs to nice **white sand beach of Hodge Bay**. Some **Manchineel Trees** in a gut just west of the wooden stairs have trunks over 1 foot in diameter.

<0.25>

6.0 (5.7) Inland is paved Solitude Valley Road. Starting about 50 yards past the intersection, the Solitude Valley Road is lined for the first 0.1 mile with **White Manjack**, a common island tree of open areas. It characteristically has several trunks, deep green, leathery leaves, and a spreading crown. It is covered with papery white, flared-trumpet flowers in summer, then with translucent white, marble-sized fruits during the fall. The fruit is a favorite food of the Scaly-naped Pigeon.

-At 15 yards further is Solitude Country Store, the only grocery on the east end of the island. M-F 8am-6:30pm; Sa,Su 8:30am-6pm. The **watermill of Estate Solitude** is in the valley N of the store. The private **greathouse and mill of Estate Solitude** are on a hill E of the store.

<0.4>

6.4 (5.3) Dirt road to left or N goes 0.15 mile to Solitude Bay's **white, sandy beach**, with shade provided by **Otaheita Trees**. A sign near the intersection says "Homesites by the Sea." One can walk from here west to Pull Point and east to Tague Bay. Wear hiking shoes and be prepared for some wading.

-Ahead to E 50 yards is a dirt road inland to **mill of Estate Cotton Valley**

0.0 Junction with Route 82 at mile 6.4(5.3)

0.6 Dirt road to left or NE. Turn onto this road.

0.7 **Mill of Estate Cotton Valley**. The **view of the northeastern coast** from this peaceful setting is unexcelled. The mill is in good shape and unmarred by modern "improvements". Sweeping from west to east, left to right:

-Edge of Coakley Bay Condominium

-Candle Reef, the sand islet offshore

-End of Tague Bay Reef with a light buoy

-Buck Island to the north

<0.3>

6.7 (5.0) Dirt road seaward to Yellowcliff Bay. The beach is mixed sand and gravel. Just offshore and about 100 yards west is

Including East End Road, Route 82, Southshore Road, Route 60, and Southside Road, Route 62

East End Road, Route 82

W-E (E-W)

a shallow reef with sponges and many other invertebrates, a nice snorkeling area.

-One can walk from here to Pull Point on the W and Reef Beach in Tague Bay on the E. The hike requires some wading and shoes to get over rocks.

<0.1>

6.8 (4.9) Paved road inland with a fire station on the SW corner. The **mill of Estate Cotton Valley** is on a hill S of the fire station.

<0.55>

7.35 (4.35) Pullout seaward for **view of Buck Island**

<0.35>

7.7 (4.0) Pullout seaward on the curve around Tague Point for **view of Buck Island and Tague Bay**. The pier to the SE is for Duggan's Reef, a restaurant and beach area for the Reef Condominium on the hills to the south. The pier further east is at the St. Croix Yacht Club. Romney Point forms the eastern boundary of the mooring area, and Cottongarden Point is the one in the east beyond. The castle on Clindinen Hill is private, a greathouse of the 1980s.

-A path at the western end of the guard rail leads through grass down to shore at Tague Point. The reinforcement bar sticking up on the reef out to sea and a little east from this point marks a scar through which one can **snorkel** across the reef. A boat ran aground on the reef, and the scar was made by dragging the boat across into the bay. The channel between Tague Pt. and the reef has substantial boat traffic. If you decide to snorkel out to the reef here, to do it safely you should tow a buoyed diver flag so that boat pilots will be able to see and avoid you. A snorkel across a barrier reef is dramatic, because you will see what immense structures reefs are, and built all of animal skeletons! At a leisurely pace it takes several minutes to cross. You will also see a change of types of fishes from the protected backreef to the wave-washed forereef and a change in coral types, with fire coral predominating on the high part of the reef.

<0.4>

8.1 (3.6) Paved road inland into Reef Condominium (private), seaward to Duggan's Reef Restaurant and pier. **White sand beach** here is very popular. The beach house (Reef Beach Watersports, telephone: 773-9088) **rents snorkel gear, windsurfers, and sailboats and charters trips for Buck Island, fishing, and SCUBA diving**. One can walk from this beach west to Pull Point. The walk requires shoes to get over rocky areas and some wading.

<0.1>

EAST END TOUR
Including East End Road, Route 82, Southshore Road, Route 60, and Southside Road, Route 62
East End Road, Route 82

W-E (E-W)

8.2 (3.5) Paved road to right or S circles the Reef Condominium to a view from the slopes of Mt. Tallin. The Reef Convenience Store on the SW corner sells items that vacationers might need. It is open every day, 9:30am-7pm.

Road around the Reef Condominium to Mt. Tallin (E-W):

- 0.0 Junction with Route 82 at mile 8.2(3.5)
- 0.2 **View of Tague Bay and Buck Island**
- 0.35 Fork. Make a right or W turn.
- 0.8 Junction. Go right onto pavement.
- 1.2 At a three-way fork, go left uphill.
- 1.45 Cul-de-sac turn around. **View of Tague Bay**, with the mooring area of St. Croix Yacht Club. Romney Point and Cottongarden Point beyond project out into Tague Bay. The two peaks with conical outlines are the Goat Hills.

<0.6>

8.8 (2.9) **West Indies Laboratory** of Fairleigh Dickinson University on the inland side. This is a teaching and research laboratory for scientific work on St. Croix and the Caribbean. A schedule of classes and other information is available from the office (first building on the left after the entrance). The next main building on the left has **exhibits** in the central breezeway.

<0.1>

8.9 (2.8) Dirt road seaward leads to the sandy beach near Romney Point. **Snorkeling** is over a shallow grassbed and rocks.

<0.5>

9.4 (2.3) Junction with the end of Southside Road, Route 60 coming in from the right or S. The East End Tour continues straight or E to East Point, but then doubles back to this point and resumes travel along the southeastern coast on Route 60. [See p. 159, mile 0.0(14.3) for Route 60.]
-About 15 yards from this junction on the seaward side of the road is a brick pillar, which marks one of the paths down to the beach at Smuggler's Cove. This is a **white sand beach** with **snorkeling over extensive, lush grass beds** of Knight Bay (also called Cistern Bay) offshore. There is no surf, since the area is protected by the barrier reef of Tague Bay about 0.5 mile offshore.

<0.1>

9.5 (2.2) Dirt road seaward to Smuggler's Cove **picnic area** with tables. The road is usually very rough, so it is best to park on the main road and walk down.
-Route 82 beyond this point provides overlooks of the patch reefs of Knight Bay, most easily visible in afternoon light with polarized sunglasses. Their corals stand out as fairly circular, golden brown areas surrounded by the deep

W-E (E-W)

green color of water over the grass bed. [See p. 160 for more about patch reefs.]

<0.6>

10.1 (1.6) Entrance to Cramer Park on the seaward side. This is a **popular beach and recreation area for Cottongarden Bay. Camping** is allowed. No fresh water is provided. The flush toilets sometimes work. The beach is a beautiful white coral sand, and the nearby snorkeling is over a grassbed, bottles, and cans. Patch reefs are accessible off Cottongarden Point, which forms the rocky wall that bounds Cramer Beach on the east.

<0.1>

10.2 (1.5) Pavement ends. This is the beginning of one of St. Croix's natural treasures, the **Fairleigh Dickinson Territorial Park**. The entire eastern end of the island is included. The forest here is typical of the dry scrub for this end of the island, which receives about 20 inches of rainfall annually. Although some may perceive it as bleak, it supports several species of cactus, including the wonderful Turk's Cap, and several locally endangered plant species. Orchids and bromeliads find a home in the cactus as well as on other plants. This is where the **Iguana Lizards** (up to six feet long including tail) are most abundant on St. Croix. You must count yourself extremely lucky if you see one. They are not common. The **beaches are pristine white sand, often deserted**, and the **snorkeling is marvelous.**

-A dirt road inland that is about 50 yards east of the pavement end, leads to the top of Goat Peak (also Sugarloaf Hill, 672 feet), the furthest east of the two Goat Hills. Access on this road may be restricted. If so, you may be able to obtain permission to walk in. From the top are **views of the north and south shores**. It affords a good look at the east end of the island and its southeastern beaches. On crystal clear days, in addition to the northern Virgin Islands, one can see Saba, a high volcanic island about 90 miles to the east. The lower western peak of the Goat Hills is Hughes Hill, 615 feet.

<0.2>

10.4 (1.3) Dirt road seaward goes to Cottongarden Point. This road is 0.15 mile long and usually in rough condition. It is best to park on the peak (less than 0.05 mile in) and walk the rest of the way. A path through tall grass leads from the road end down to the shoreline.

-Just slightly east of this junction one comes to a **view of Boiler Bay**, described at the next stop. The nearshore "boilers" are easily visible from this point in afternoon light.

<0.15>

EAST END TOUR
Including East End Road, Route 82, Southshore Road, Route 60, and Southside Road, Route 62
East End Road, Route 82

W-E (E-W)
10.55 (1.15) Two trees on the seaward side of the road are **Fustick** (also San Bartolomé). The bark is grooved, grayish-white. The showy, tubular, orange flowers and pear-shaped fruits are usually borne in spring and summer. Leaves are 5 inches long with orange veins, and they have a coarse feel like sandpaper on the tops. This tree is found only in Puerto Rico and the Virgin Islands.

Fig. 48. Dense coral gardens of Bluegut Bay.

-A dirt road seaward, about 30 yards east of the Fustick Trees, goes to the **beach of Boiler Bay** (also called Brown Bay, Whale Bay). This Bay is named for the large rocks, 7-10 yards offshore, which are covered with algae. They usually have breaking waves except in very calm seas. These boilers are algal reefs, made by rock-depositing algae rather than by corals. **Snorkel** around the boilers. They have many overhangs and caves that harbor a diversity of fishes. Their surfaces are pocked with the burrows of the **Rock-boring Urchins**, which graze on both the algae growing near their burrows and drift algae. Feed a piece of algae to an urchin as though it were a serendipitous drift catch. Touch the algae to the urchin's spines and observe what happens. After several minutes, you may see the urchin transfer the algae to its mouth on the underside of its shell. Further offshore, Boiler Bay has **patch reefs in seagrass beds and a barrier reef**.

EAST END TOUR
Including East End Road, Route 82, Southshore Road, Route 60, and Southside Road, Route 62

East End Road, Route 82

W-E (E-W)
<0.45>

11.0 (0.7) Dirt road seaward that forks immediately goes to **beaches.**
-The left fork goes 0.1 mile to the eastern end of Boiler Bay and continues to Whale Point at 0.15 mile. You can climb up to Lamb Peak (68 feet), the rise just behind you, for a view. Below the Point in the water are two turquoise lanes of sand. These are called "The Colorado River," because the current is nearly always flowing swiftly along them from Bluegut Bay (on the right or E) to Boiler Bay (on the left or W). Some people like to start in Bluegut Bay and ride the current to Boiler Bay, walking back over this road.
-The right fork goes 0.1 mile to the western end of Bluegut Bay (also called Lamb Bay).

<0.15>

11.15 (0.55) Dirt road seaward to the eastern end of Bluegut Bay (Lamb Bay). This bay provides **very nice snorkeling among hard corals.** The far eastern part of Bluegut Bay can be reached by climbing about 200 yards along the shoreline rocks, and has a separate name, François Bay.
-Beyond this junction is a steep rough section of road, but it is only 0.35 mile long and almost always passable by two-wheel drive vehicles <u>with high clearance</u>.

Fig. 49. Turk's Cap Cactus on East End show many variations in their top-knots--single fez; multiple fezzes; elevated fez; and cristate.

EAST END TOUR

Including East End Road, Route 82, Southshore Road, Route 60, and Southside Road, Route 62

East End Road, Route 82

W-E (E-W)
<0.35>

11.5 (0.2) Dirt road to right or S goes to three **lovely, isolated, white sand beaches**. From this junction you can see the effect of nearly constant windshear from the tradewinds on the vegetation of the hill Morne Rond. The dirt road is usually impassable even by four-wheel drive vehicles, so it is best to leave vehicles at the junction and **hike to the beaches**. Distances from here to the beaches are: nearest, 0.25 mile; second beach, 0.5 mile; third beach, 1.25 mile. You will need shoes, and long pants are helpful for moving through the brush of Isaac Point if you are going all the way to Jack Bay, the furthest beach. The road forks at 0.05 mile from the junction. Both forks run 0.2 mile to the first beach at East End Bay.

-East End Bay (also called Windward Bay) has a fringing reef of hard corals, easy to see and wade among wearing foot protection and long pants. The white sand beach is excellent for **hunting for shells, seeds, and jetsam**. Plentiful here are the beautiful, delicate, 1-inch, white shell coils of the Common Spirula, a squid-like animal. Drift seeds are abundant. These are from plants that have seaworthy seeds which can travel a good distance by floating, and then, unscathed by the journey, sprout and colonize where they wash ashore. Some of the most common are the little green apples of the Manchineel. One of the best-known drift seeds is the Coconut.

Fig. 50. Forest of Dildo (tall columnar cactus) and Padded Tree Cactus on the walk between Isaac and Jack Bays.

East End Beaches (continued):
-The next beach is Isaac Bay, a **splendid half-mile extent of white sand beach**. To get there from East End Bay Beach walk SW up over Point Cudejarre. The trail is up the natural gullies in the bedrock at the isthmus of the point. Be sure to examine the **Turk's Cap Cactus** that are so abundant on this trail. They are the short, barrel cactus with the red tops. The tall, columnar cactus are called **Dildo** (honest). Isaac Bay has a big reef area and good **snorkeling**.
-The third beach is Jack Bay. To get there one must walk over Isaac Point, but here the paths are overgrown because fewer people walk beyond Isaac Bay. An old trail leads up over the isthmus from about 100 yards northeast of the SW end of Isaac Beach, but it is badly overgrown on the Jack Bay side. An easier route is to climb up the rocks, again using the natural gullies at the end of the beach. On Isaac Point, stay to paths that run near the cliff edge, just outside the main clumps of columnar cactus. From the eastern end of the Point as one rounds the bend into Jack Bay one can look down onto a very nice **algal ridge**, a reef formed by rock-depositing algae rather than coral. It has a dark green color, rather than the golden brown of coral, and waves are usually breaking on it. Find natural openings in the brush and trees that border the shoreline to get to the **white sand beach**. **Snorkeling** here is over an excellent grass bed.

<0.2>
11.7 (0.0) Turnaround on the peak of East Point (Kid Peak, 226 feet). The view from the top on a clear day includes the northern Virgin Islands, 40-60 miles away, and Saba, a volcanic island 2828 feet high, 90 miles to the east. The point below is named Point Udall after a former Secretary of the Interior. This is the most eastern point of land belonging to the United States in the Western Hemisphere. (In the Eastern Hemisphere one of the Aleutian Islands has this distinction.) You can **hike** down to the sea and East Point along any of various paths that start just over the railing. One of the easier routes is along the ridge line. It starts 7 guard rail supports south of the "End" sign. Most people do not make the short walk out to the tip. They miss a lot. Those waves are puny and silent from up on the road. But they are exploding on the rocky point. Take your camera. The barrel cactus with a red top are **Turk's Cap Cactus**. As you drop down the slope, watch how the vegetation diminishes in height and abundance until finally it loses the battle against constant wind, sea spray, and rockfall. Near the shoreline stay off of the wet rocks, for this is where the highest waves are hitting. The

W-E (E-W)

highest waves often come in intervals several minutes apart, so the wet rocks are the best clue to where the danger zone begins. Where the waves are breaking at the moment is not a good clue. The uptilted rock layers belong to the Caledonia Formation, 80 million years old, part of the oldest exposed rocks on St. Croix.

Fig. 51. The horseshoe reefs in the water are deep green algal ridges off Isaac Point, formed by rock-depositing algae and not by coral. This type of reef is always in a zone of intense wave action.

SOUTHSHORE ROAD, ROUTE 60

(also South Side Road)

This part of the East End Tour begins at mile 9.4(2.3) [p. 152] of Route 82, the Smuggler's Cove Corner. From East Point you will have to retrace your steps for 2.3 miles to this junction to continue the tour. This road provides access to very nice beaches along the south shore, good snorkeling areas among the coral reefs, some scenic viewpoints for the south shore, and a wonderful birdwatching area at Great Pond. In 6.3 miles near Great Pond, there is a connection via Route 624 to Route 62, and the tour finishes on Route 62. Note that you have changed direction, and are now going E-W towards Christiansted for this part of the tour.

E-W (W-E)

0.0 (14.3) Start of Route 60 at junction with Route 82 [p. 152, mile 9.4(2.3)]. A pleasant beach area, Smuggler's Cove, is on the NE corner of this intersection.

<0.3>

0.3 (14.0) Junction with paved road. Turn right or W onto N. Grapetree Bay Hill Road to **viewpoints from Romney Hill for Tague and Knight Bays** on the north shore.

Grapetree Bay Hill Road to Romney Hill (E-W):

0.0 Junction with Route 60 at mile 0.3(14.0)

0.2 Fork. Turn right or N onto No. Grapetree Bay Circle for **viewpoints**.

0.35 A path on the left goes to a **lookout from top of Romney Hill (378 feet) over Tague Bay**. It is located just 10 yards before the pavement ends. The walk to the peak is about 50 yards. Proceeding from the distant west:
-Peninsula of Estate Judith's Fancy
-Inland peaks, moving from the shoreline towards the interior: Saltriver Mountain with two peaks, Clairmont Peak (866 feet) and Windsor Peak (885 feet), Mt. Eagle (1165 feet), Blue Mountain (with antennae, 1096 feet), Seven Hills Range with the furthest and highest Peak Seven (860 feet); Bee Peak of Thunderhole Hill, a nearby and seemingly very high dome (600 feet)
-On the shoreline going from left to right, west to east: Green Cay, Pull Point, Candle Reef (an islet out to sea), the western end of Tague Bay Reef, Tague Point, Duggan's Reef pier, the moorings and dock of the St. Croix Yacht Club
-Behind you to the east are the Goat Hills, with Goat Peak the furthest and highest (660 feet).

0.45 A nice **lookout over Knight Bay**. Knight Bay is formed by Cottongarden Point on the east, the Tague Bay barrier reef on the seaward side, and Romney

Grapetree Bay Hill Road, E-W (continued)

Point on the west. The two high peaks in the east are the Goat Hills. Just west of Cottongarden Point in Knight Bay one can see golden circles of **patch reefs** surrounded by the deep green of the turtle grass beds. (This is best seen in afternoon light with polarized sunglasses.) There are about 18 patch reefs visible from this point. The patch reefs are underlain by uptilted areas in the bedrock. One theory for patch reef formation is that they were small hills in the bay. Corals were able to maintain coverage in these elevated areas, while coral of the lower areas was rapidly, or, at least eventually, covered over by sediment. This left the patch reefs as little islands of coral surrounded by the grassbed.

-Channel Rock Reef is the dark-colored area in the Buck Island Channel seaward of the barrier of Tague Bay Reef. The sandy indent (light turquoise blue area) in the Channel Rock Reef is frequented by porpoises.

<0.25>

Route 60 (continued)

E-W (W-E)

0.55 (13.75) Paved road to left or E into residential area of S. Estate Grapetree Bay and to Grapetree Bay Hotel. The western end of the Grapetree Beach is usually very nice **white sand beach** with some surf.

Grapetree Bay area (W-E):

0.0 Junction with Route 60 at mile 0.55(13.75)

0.15 Paved road to right or W leads immediately to a vacant field on the seaward side. One can walk through the field to enter Grapetree Bay beach for snorkeling. This bay once had a very nice stand of **Elkhorn Coral**, an important reef-builder. The coral has massive, branched colonies of golden color, often two yards or more in diameter. The colony form is similar to moose antlers. It is one of the fastest growing corals and contributes substantially to the framework of shallow reefs, that is, much of the dead coral to which the bases of live corals are attached is this species. In the past several years St. Croix and many other areas in the Caribbean have experienced an epidemic of a disease that attacks Elkhorn Coral. It is called "White-band Disease", and the causative agent is unknown. The disease starts as a white band, usually near the base of a coral branch. The band grows outward towards the end, leaving the coral dead (without golden-brown

Grapetree Bay area, W-E (continued)

colored tissue) in its wake. You will see the remnant pieces of Elkhorn Coral lying all over the bottom here, like scattered giant jigsaw puzzle pieces, but here and there are colonies which have survived the disease, and other small ones which may be new. The course of this disease is certainly a cause for concern, for the character of St Croix's reefs has changed dramatically with the widespread die-off of the dominant member of the shallow-water corals. It is unknown whether the reefs will recover. It could be that this is a cyclic phenomenon that has occurred previously, but which was unobserved before the widespread interest in snorkeling and SCUBA diving began in the 1940s.

0.2 Entrance to Grapetree Bay Hotel. The hotel's beach was created by dredging sand from Turner Hole, a mile to the W. The groin construction to retain the sand cost $100,000, but Nature is not so inexpensively thwarted. The prevailing currents periodically bare the beach to its natural, rocky base, and the fortunes of the hotel seem to come and go along with the sand. The walk along the beach towards Hughes Point on the east is pleasant, but the cliffs are too steep for one to get all the way around to Jack Bay.

0.25 End of road 0.5 mile from Grapetree Bay Hotel at hiking trail to the **fine, white-sand beach of Jack Bay** [see p. 157, mile 11.5(0.2)]. The trail is very short, just down the cliffside to the beach. If parking is not allowed along this road, park at the hotel.

<0.15>

Southshore Road, Route 60 (continued)

E-W (W-E)

0.7 (13.6) Grassy road between lots 27 and 26 goes to a very nice **white sand beach of Grapetree Bay called Meggie**. At the seaward end of the path on the left or E is a large-trunked **Water Mampoo Tree** (also called Loblolly). It is characterized by large nearly round leaves, 3-10 inches long, and a gray smooth trunk. When the wind is right, this is a good body-surfing area. Just offshore is a nice snorkeling area among large rocks with caves and fissures. These rocks are between 50 and 100 yards offshore of the second house west of the groin wall.

<0.45>

1.15 (13.15) "Rocky Hill", road inland uphill. At 0.1 mile there is a good overlook of Turner Hole from the east with morning light.

<0.3>

Including East End Road, Route 82,
Southshore Road, Route 60, and Southside Road, Route 62

Southshore Road, Route 60

E-W (W-E)

1.45 (12.85) Dirt road seaward goes to sandy beach of Turner Hole (also called Thunder Hole). Fishermen launch boats here.

<0.2>

1.65 (12.65) Grapetree Beach Hotel. An imitation mill is at the main entrance. There is a nice **sandy beach, seagrass bed lagoon** (Turner Hole) and **offshore reef**. One entry for **snorkeling** on the reef of Turner Hole is to walk west from the hotel about 200 yards along the beach. One passes a dense stand of **Seagrape Trees** for the first 100 yards and then beside a **coconut grove** for the last part of the walk. Look for a sand area (light blue among the deep green of the grass bed) for easy entry.

<0.2>

1.85 (12.45) Dirt trail seaward goes to **white sand beach** of Turner Hole opposite the "Welcome/Leaving Grapetree Beach Hotel" sign. A chain bars the entrance to vehicles.

<0.35>

2.2 (12.1) A good **view of the Turner Hole area** just before rounding a sharp curve. A narrow pullout is on the seaward side. The curve is blind. Eastbound travelers (W-E) should continue past the pullout, turn around where they can see the oncoming traffic, and return to the pullout. If you are planning to snorkel in Turner Hole, examine the water from this vantage point to get an idea of where you may want to enter the water and where you will want to go among the corals.

-For a **really good view of the southeastern beaches**, you can climb up Pentheny Peak (325 feet) behind you. Start uphill about 30 yards NE from the apex of the curve, just as the paved road starts to drop, keeping the main group of columnar cactus on your right. An indistinct path through tall grass goes up the ridgeline to the top. Along the way is a fine view to the west. In the distance and coming towards Pentheny: Mt. Fancy, with Mt. Fancy Saltpond below; Robin Bay with its algal reef barrier; Rod Bay; and Grass Point. To the east and coming towards Penthany: Isaac Point in the distance with part of Jack Bay visible; Hughes Point cutting off the view of the nearer part of Jack Bay and marking the eastern end of Grapetree Bay; Grapetree Point which marks the eastern end of Turner Hole and the cut through the reef out to sea; Turner Hole; and Grass Point. The high hill inland above Pentheny is Thunderhole Hill with Bee Peak, 600 feet.

<0.25>

2.45 (11.85) Dirt road seaward goes 0.3 mile to Grass Point. Four-wheel drive may be necessary to reach the top of the knoll, but two-wheel drive vehicles simply park at its base, and one can then walk less than 0.05 mile to the top. Grass Point

E-W (W-E)

is usually wave-buffetted and scenic because of crashing waves. The **tidepools** are inaccessible except in calm seas. If the sea is not rough, this is an interesting area to **dive or snorkel**, since the nearly vertically dipping bedrock forms fissures, crevices, and caves underwater near the Point. The best entry and exit point is to the right or W in Rod Bay. Entry is done by walking over the cobble bar. Entry and exit on the east side (Turner Hole) is best attempted only in calm conditions.

Fig. 52. A riot of coral in Turner Hole, viewed from Pentheny.

<0.05>
2.5 (11.8) At green mileage sign "5", a dirt road to shore goes 0.1 mile to Rod Bay (or Red Bay). This part east of Carty Point is sometimes called Carty Bay. Seagrass bed and extensive coral patches make this an interesting **snorkeling area**. The barrier reef is fairly close to shore.

<0.4>
2.9 (11.4) Dirt road to shore goes 0.1 mile to Rod Bay. A few Manchineel Trees provide shade at the turnaround area on Carty Point.

<0.1>
3.0 (11.3) Dirt road shoreward at the green mileage marker "4.5". This western part of Rod Bay is sometimes called Woodcottage Bay. A nice forest of **Manchineel Trees** provides welcome

E-W (W-E)

shade, but be careful of the trees. Their little green apple fruits are poisonous.

<0.15>

3.15 (11.15) Dirt road on shore side. A fork at 0.05 mile at the top of Fanny's Fancy Point provides a nice **view east into Rod Bay with Grass Point in the distance**. Veer right at the fork to continue to end at 0.1 mile. This last portion may require four-wheel drive. Directly at the base of the hill is a sizeable patch reef, detectable by its dark oval shape, about 150 yards long, running parallel to shore. A reef usually visible due to breaking waves, just west and offshore about 250 yards, is an algal ridge.

<0.35>

3.5 (10.8) A dirt road goes seaward to the east part of Robin Bay. A seagrass bed is directly offshore. The offshore reef, a large algal ridge, is usually obvious due to breaking waves.

<0.15>

3.65 (10.65) Dirt road seaward to Madsen Point. The part of Robin Bay to the west is sometimes called Cottongrove Bay. The road is 0.2 mile long, ending in a loop.

<0.2>

3.85 (10.45) Four-wheel drive road seaward to Cottongrove Strand on Robin Bay and turtle grass beds.

<0.2>

4.05 (10.25) Dirt road to shore (Cottongrove Strand). At 0.1 mile it passes Mt. Fancy Saltpond, a **birdwatching area**. At 0.15 mile the road turns along the shore. **Snorkeling** here in Robin Bay is over turtle grass beds.

<0.2>

4.25 (10.05) Inland, a private house around **mill of Estate Cotton Grove**

<0.45>

4.7 (9.6) A sharp, blind curve at a ridgeline crossing of Mt. Fancy has a pullout on the shoreward side. To stop for a view when traveling E, turn around further down the road where the oncoming traffic is visible. Proceeding from the sea to land, this **scenic view of the Great Pond area** with Milord Point on the west looks over Great Pond Reef, Great Pond Bay with numerous patch reefs [see p. 160], Great Pond berm and mangrove, Great Pond Saltpond and salt flats.

<0.3>

5.0 (9.3) Dirt road on shore side goes 0.2 mile to a parking area. Walk along a path through dry forest brush, about 0.1 mile from the parking area to the shoreline of Great Pond Bay. The path ends at the beach near the ruins of a small concrete pier. A nice seagrass bed is just offshore.

<0.4>

5.4 (8.9) Dirt road on shore side goes to **Camp Arawak**, built amid **ruins of Estate Great Pond**. **St. Croix Agave**, known only

E-W (W-E)

from St. Croix, lines the western side of the entrance road. The beach area here is silty sand with grass. To use the area, register in the office in the reconstructed old greathouse. Day use of the beach is free. Camping in your own tents is $5/day/campsite. Fresh water and pit toilets are provided. Rental of canoes and some diving equipment is available. Address: Arawak Program Inc. (Camp Arawak), Box 129, Christiansted, V.I. 00820. Telephone: 809-773-3944.

<0.15>

5.55 (8.75) Route 60 veers right or N at this point. The dirt road to the left or S is one of the **main access points for the shoreline of Great Pond Bay and the shoreline mangroves of Great Pond**. It is a launch spot for fishermen.

Fig. 53. A young black mangrove tree with radiating root lines of pneumatophores.

Great Pond inlet (N-S):

0.0 Junction with Route 60 at mile 5.55(8.75). The road runs through a nice **stand of Black Mangrove Trees** along the western end of the salt flats. This is a good **birdwatching area.**

0.15 Shoreline. Turn right or W.

0.25 End of road at mouth of inlet to Great Pond. The inlet is lined with **Red Mangrove Trees.**

Southshore Road, Route 60

E-W (W-E)

<0.05>

5.6 (8.7) Dirt road to left or W to salt flats of Great Pond

<0.1>

5.7 (8.6) Dirt road to left or W to salt flats of Great Pond. Use caution and park on firm ground. After rains this area is usually impassable.

<0.2>

5.9 (8.4) Dirt road to left or W is the **main route to the saltpond area of Great Pond Bay for birdwatching**. Use caution on these roads, as they can be very muddy. Note that the one-way distance for walking in is only 1 mile.

Great Pond perimeter, (E-W):

0.0 Junction with Route 60 at mile 5.9(8.4)

0.15 Turn right. After 50 yards, turn left or W along the fence line at the N end of the Great Pond area and drive along the northern edge of the saltpond.
-A dirt road off Route 60 to left or W about 15 yards N of the road above goes to this fenceline road. Although very rough, that road can sometimes be used when the main entrance road is too muddy.

0.4 Turn left or S through a broken fence line and travel along the western side of the saltpond.

1.05 The road ends at a cul-de-sac in the southwestern corner of the saltpond. Park here and walk.

-This southern end of the Great Pond and the mangrove-lined shoreline is a **fine birdwatching area**. On a typical walk you should see Brown Pelicans, Magnificent Frigatebirds, Semi-palmated Plovers, Wilson's Plovers, Black-bellied Plovers, Black-necked Stilts, Great Egrets, Great Blue Herons, Green-backed Herons, Yellow Warblers, White-crowned Pigeons, Black-whiskered Vireos, Caribbean Elaenia, and several kinds of terns. Least Terns sometimes breed near the end of the road. In the winter, migrants make the list of typical birds much longer.

-To take the **Great Pond Nature Trail**, walk east along the vegetation line on the Great Pond barrier berm for about 70 yards until you come to a cleared path into the vegetation on the berm. It is about 3 yards wide. The path divides almost immediately.

-The western narrow part leads straight south to the shoreline of Great Pond Bay. Walking west along the beach about 600 yards, one comes to the trailhead at the Boy Scout Camp, mentioned below at mile 8.15(6.15). There are some very large **White Mangrove Trees** near the shore at the trailhead.

-The eastern three-yard-wide section of the trail goes eastward along the top of the berm through a shady dry

Great Pond area (continued)
lowland forest dominated by Manchineel Trees. About 600 yards to the east is a clearing with two large Tamarind Trees among the Manchineels. Here walk north towards the saltpond to continue on the Great Pond Nature Trail. The Trail runs just along the shoreline of the saltpond, with a line of Black Mangroves providing good cover for birdwatching. Cross a peninsular extension of the berm to continue on the Trail along the saltpond. The trail ends at the inlet to the saltpond, which is lined with Red Mangrove Trees. If you want to return by way of the shoreline of Great Pond Bay, you will have to wade in silty mud for part of the distance.

<0.4>

Southshore Road, Route 60 (continued)
E-W (W-E)

6.3 (8.0) Junction with Route 624 on the left or W. Turn left or W onto Route 624 to continue the East End Tour. Route 60 continues straight or N 1.5 miles to Route 82 [p. 146, mile 3.35(8.35)] and provides access to **scenic viewpoints** in the Seven Hills Range. This route is described in "Tipperary Road, Route 60" [p. 178, mile 1.5(0.0)].

SOUTHSIDE ROAD, ROUTE 62

(Also Southshore Road)

<0.8>

7.1 (7.2) Junction of Route 624 and Route 62. To continue the East End Tour turn left or S at this junction. Route 62 continues north 1.9 miles to Route 82. A short distance away on this northern leg is an **excellent viewpoint of the Great Pond Area and the middle of the island** at the top of Maria Hill, 387 feet.

Lowry Hill Road, Route 62 (northern leg to Route 82):
S-N (N-S)

0.0 (1.9) Junction with East End Tour, Route 624, at mile 6.9(7.0)

<0.7>

0.7 (1.2) Dirt road to right or E goes to **viewpoint from Maria Hill**, 387 feet.

0.0 Junction with Route 62 at mile 0.7(1.2)
0.5 Fork; turn left.
0.6 Spiral concrete structure at top of Maria Hill. In view from here go clockwise, starting with the high hill with antennae that is just west of Maria Hill:
-Lang Peak (760 feet), just right of peak with numerous antennae

Southside Road, Route 62

View from Maria Hill (continued)

-Flat top of Recovery Peak (845 feet) with short tower

-Blue Mountain (1096 feet) in distance with antennae and dishes

-Mount Eagle (1165 feet), next right high peak

-Windsor Peak (885 feet), second right peak from Mt. Eagle

-Peninsula of Estate Judith's Fancy

-Christiansted Harbor

-Altona Lagoon beneath radio tower

-Mill of Estate Butzberg on a low hill inland from Altona

-Green Cay, and a small part of Southgate Saltpond inland

-**Mills of Estate Southgate**, almost hidden by hills

-**Mill and ruins of Estate Green Cay** on a hill near the northern coast

-Buck Island, out to sea, 329 feet high

-Highest peak to east is Peak Seven of the Seven Hills (866 feet)

-**Mill of Estate Sight** on a low hill right of Peak Seven

-**Great Pond area**: The saltpond and salt flats, barrier berm, Great Pond Bay and its barrier coral reef, the pass into Great Pond Bay adjacent to Milord Point

-**Mill of Estate Petronella** to the south, in line with Milord Point. Nugent Peak (185 feet) is on Milord Point.

Fig. 54. Great Pond area, viewed from Greatpond Hill Road. The Salt Pond is in the center, the berm is between the pond and the ocean, and the barrier reef forming the lagoon is evidenced by the line of surf.

Lowry Hill Road (continued)
S-N (N-S)
<0.75>
1.45 (0.45) **Mill of Estate Boetzberg** (private) on the left or NE
<0.45>
1.9 (0.0) Junction with Route 82 [p. 142, mile 1.35(10.35)].

<0.3>
Southside Road, Route 62 (continuation of East End Tour)
E-W (W-E)
7.4 (6.9) **Mill of Estate Petronella** on hill to right or W with Castle Nugent Farms built around the ruins.
<0.75>
8.15 (6.15) Veer to right or W. At 25 yards further W is the Boy Scout Camp among ruins of **Estate Fareham**. This is the only **campground** on the island with tent rentals. Day use fee is $1. Camping is allowed with your own tent, $10/day/person, or tent and mattress are provided for $8 per day more. Kitchen facilities include refrigerators. There are showers, lavatories, electrical outlets, and flush and pit toilets, and boats. Address: Boy Scouts, Box 1353, Frederiksted, V.I. 00840, ATTN: Camp Ranger. Telephone: 809-773-1733.
-The **Great Pond Nature Trail** starts at the eastern end of this camp. Drive through the gate and turn left immediately. The dirt road curves towards the sea and ends at about 0.2 mile. Walk from here towards some **very large White Mangrove Trees**. The trail starts just east of these trees, running through some Manchineel groves and then out onto the seaward beach of the Great Pond berm. About 600 yards east of the Boy Scout Camp, look for a trail leading inland to the top of the berm. For a description of the trail from that point on, see "Great Pond Nature Trail" [p. 166].
<0.35>
8.5 (5.8) **Mill and ruins of Estate Fareham** on shoreward side
<0.3>
8.8 (5.5) Private **greathouse of Estate Nugent**, inland
<0.6>
9.4 (4.9) Spring Bay is to the south of the road. The exposed reef-like structure is called a "barrier rampart". It is made of coral skeletons that were turned up and accumulated here during the close passage of Hurricane Allen of 1979. The sea level was about 5 feet higher than normal during that period of very rough weather. Since no barrier ramparts were on St. Croix before Hurricane Allen, it may be that erosion wears down these structures in a few decades.

<1.0>

Including East End Road, Route 82,
Southshore Road, Route 60, and Southside Road, Route 62

Southside Road, Route 62

E-W (W-E)

10.4 (3.9) Junction with Springgut Road, Route 85. This rough dirt road goes 2.5 miles north to Route 82 [p. 139, mile 0.65(11.05)].

<0.05>

10.45 (3.85) Private **ruins and chimney of Estate Longford** on left or S. Throughout St. Croix and especially in the pastures along Route 62 you will see cattle with a rich mahogany color. This is the **Senepol**, a hybrid of the African N'Dama and the European Red Poll. The Senepol, developed on St. Croix, combines the N'Dama tolerance for insects and tropical heat with the gentleness and good meat of the Red Poll.

Fig. 55. The Cattle Egret is often seen feeding near cattle and other large, earth-shuffling objects like tractors. They eat insects and other small animals that the cattle scare up. This bird has a very interesting history of recently spreading in the New World. When collectors first came to America following Columbus, they obtained no specimens of this now very abundant bird, probably because it did not exist in the Americas. Presumably, individuals blew over in the equatorial winds from Africa from time to time, but never established themselves. Although there were vast herds of grazing animals in North America (buffalo, antelope), vagrants from Africa would have first had to land and survive in the Caribbean area where there were no grazing ungulate herds. Colonists, of course, changed that pattern by bringing livestock with them. Egrets then arriving when herds of livestock were present, established themselves. The egrets gradually spread up the Caribbean chain to Florida in the 1950's, and then to the West Coast of the United States by the 1960's. The cow in the illustration is a Senepol.

Southside Road, Route 62

E-W (W-E)
<0.4>
10.85 (3.45) Dirt road to right or N goes into a residential area with a good **south coast view**.

S-N:
- 0.0 Junction with Route 62 at mile 10.85(3.45)
- 0.4 Three-way fork. Go right or E.
- 0.7 **Viewpoint of south shore**. Sweeping left to right, E-W are Spring Bay with an offshore barrier rampart, Halfpenny Bay; Manchenil Bay with the **mill and ruins of Estate Diamond Keturah** on a low rise about halfway along the bay margin; Watchho (Vagthus Point) peninsula; and Hess Oil Refinery.

<0.4>
11.25 (3.05) Junction with dirt road. To left or S through the four pink concrete pillars leads to a popular, **very fine one mile stretch of white sandy beach**, Manchenil Bay.

Manchenil Bay Road (N-S):
- 0.0 Junction with Route 62 at mile 11.25(3.05)
- 0.65 Parking area for Manchenil Beach and Bay where the road veers to left along the shoreline. This beach is a favorite for jogging, walking, and relaxation. Offshore is a seagrass bed. The beach berm has several nice shady groves accessible by walking along the beach. The walk to Watchho (the elevated point of land on the western end of the bay) is about 1 mile. Watchho is also called Vagthus Point, "vagthus" being the Danish equivalent for "watch house."

  -A clump of endemic **St. Croix Agave** is about 70 yards west of the parking area at the margin of the dense tree growth that borders the beach. To see it, walk about 15 yards in from the water's edge. It has a very interesting propagation mode, bearing bulbils which grow into little agave plants on the flower stalk. The seeds sprout while still on the stalk, and these miniature plants then fall off and root.

  -About half-way along the beach one can cut through the beach berm to some inland sand flats. **Shore birds** feed in this area.

  -You will need hiking shoes for climbing up the cliffs of Watchho. Keep to the outside of the fence by walking along the outer margin of the point. The **White-tailed Tropicbird** breeds during the winter on Watchho and on some cliffs to the west which border Canegarden Bay.

Manchenil Bay, N-S (continued)

Fig. 56. St. Croix Agave at Manchenil Bay. The seedlings cover the flower stalk. They will fall to the sand and root.

0.75 **Beach access** to Batist Point (also Fishermen's Point) on shoreward side of the road. The bedrock dips vertically here, making an interesting snorkeling area. One can walk E along the beach from this point by wading out around the trees at the nearby houses (the last ones on the beach for miles). Go to the pillars and fence, then turn right and walk outside the fenceline through the Seagrape Trees to the beach. One can **hike** for about 2.5 miles of shoreline: Halfpenny Bay, Spring Bay, and Fareham Bay. The windward shores have lots of jetsam.

Manchenil Bay, N-S (continued)

Some nice **conch shells** are in piles left by fishermen. The lagoons are covered by turtle grass beds and are shallow enough for wading. At the eastern end of Halfpenny Bay there is a cattle fence across the beach that is easy to cross.

<0.45>
11.7 (2.6) Junction with Route 83, a paved road to right or N. Route 83 goes 1.9 miles to join Centerline, Route 70.

Route 83 (S-N):
0.0 Junction with Route 62 at mile 11.7(2.6)
0.25 Private **greathouse of Estate Granard** on left or W
1.3 Junction with Route 622. Turn left or W.
1.4 Turn right or N to continue Route 83.
-At 30 yards north of the intersection on the left or W is a house with a large **Lignumvitae Tree** surrounded by a concrete planter wall.
1.45 Freshwater pond on east side of road
2.0 Junction with Centerline, Route 70 [p. 24, mile 2.1(13.3)].

<0.2>
11.9 (2.4) **Mill and ruins of Estate Diamond Keturah** can be seen on a low hill on the left or S side of the road

<0.6>
12.5 (1.8) **Baby house ruin of Estate Canegarden** on left or S of road. [See p. 26, mile 4.65(10.75) for a description.]

<0.2>
12.7 (1.6) Private **greathouse of Estate Canegarden** on left or S

<0.85>
13.55 (0.75) Paved road. Left or W goes to access for French Pond, an **excellent bird-watching area** at Canegarden Bay. The walk to Canegarden Bay is about one mile, one-way, passing through a dense scrub forest of Acacia and other native trees. Land snails are very abundant along the way. Another access to French Pond is described on p. 53, mile 2.05(0.8).

French Pond and Canegarden Bay (E-W):
0.0 Junction with Route 62 at 13.55(0.75)
0.2 End of road at chained entrance to Southside Pentacostal Church. Park here and walk about 1 mile to Canegarden Bay area of saltponds and salt flats.

(1) Walk W along the dirt road that continues from the north boundary of the church.
(2) The road ends after about 0.25 mile in another dirt road. Turn left or S.
(3) Veer right or W at the first fork (after about 0.4 mile more).

E-W (W-E)

(4) The branches of the next fork take you to different sides of the mangrove ponds.
(a) The right or W fork goes along the northern edge of the largest saltpond. One must slip through bush to get to the edge of the pond, but several areas have brush that is not dense.
(b) The left or S fork leads to the southern part of the largest pond and seaward access. The walk to the sea passes on the western side of an abandoned house. At the shore walk down over the limestone ledge to the sand and cobble beach. From the beach one can easily reach the southern end of the large saltponds and the salt flats on the western side of the area. Birds in this area include many winter migrants that are uncommon elsewhere such as Common Snipe. This is a good area for White-crowned Pigeons and Black-crowned Night Herons. White-cheeked Pintails and Black-necked Stilts breed in the largest pond.

<0.60>

14.15 (0.15) Junction with Route 68 [p. 54, mile 2.85(0.0)].

<0.15>

14.3 (0.0) End of the East End Tour. Junction of Routes 62, 682, and 70. Route 682 is a left or W turn, and goes 0.4 mile to Route 68 [p. 53, mile 2.5(0.35)]. Centerline, Route 70, goes both right or E and straight or N [p. 25, mile 3.5(11.9)]. Turn right to go E on Centerline, Route 70, to Christiansted. Continue straight or N to go to the shopping centers and Frederiksted.

SIDE TRIP OFF EAST END TOUR

TIPPERARY ROAD, ROUTE 60

This section of Route 60 (also known as "Southgate-Great Pond Road, Cross Island Road") provides access to some nice viewpoints.

N-S (S-N)
0.0 (1.5) Junction with Route 82 [p. 146, mile 3.35(8.35)].
<0.4>
0.4 (1.1) Paved road to left or E. This road provides access to good views of the northern shore from the area of Seven Hills.

All-for-the-Better Road (W-E):
0.0 Junction with Route 60 at mile 0.4(1.0)
0.35 Fork. Make a left turn.
0.5 Three-way Fork. Take the right paved road uphill.
1.1 Fork. Take the left fork, which looks like a driveway, and leads 0.1 mile to a housesite, an excellent viewpoint until someone builds there.
The westernmost peninsula in the distance is Estate Judith's Fancy, which forms the eastern boundary of the Salt River Estuary, hiding it from view. Saltriver Mountain is west of the peninsula with two peaks, Clairmont Peak crowned by a house (866 feet), and Windsor Peak (885 feet). The two highest peaks inland comprise Eagle Ridge, with Mount Eagle (1165 feet) furthest away, and Blue Mountain with many antennae (1096 feet) further S. Long Reef is the barrier coral reef forming Christiansted Harbor. Altona Lagoon is the very large mangrove just east of Christiansted. Southgate Pond is a sizeable mangrove lying further east beside the Green Cay Marina. Green Cay is just offshore of this area. Pull Point is isolated by an isthmus, just east of Green Cay. Due N of you is Coakley Bay Saltpond, and offshore from it is a sandy islet, Candle Reef. The barrier reef of Tague Bay is over 3 miles long. It starts just east of Candle Reef at about the same distance offshore. Buck Island is the large island offshore.

<0.3>
0.7 (0.8) Paved road to left or E (just before road veers to right or W) provides **good views of the north- and southeastern shores** from high housesites in the southern part of the Seven Hills on Greatpond Hill.
-At 30 yards further south an A-frame house on the right or W has several large **Gregre Trees** on both sides of its gravel driveway.

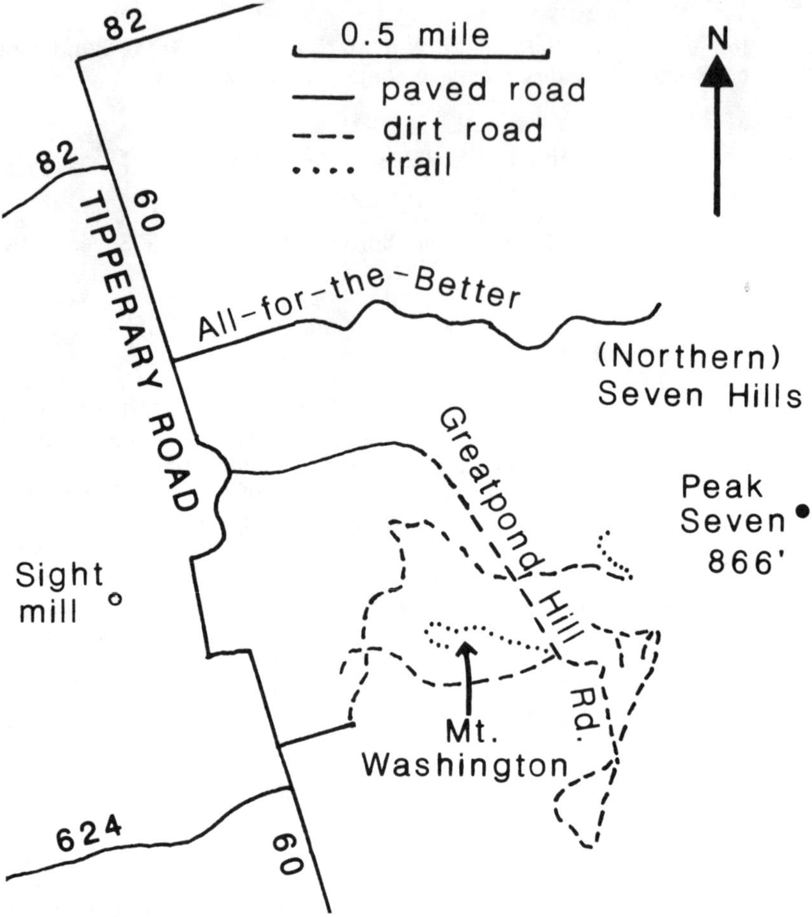

Fig. 57. Map of Tipperary Road showing side trips to All-for-the-Better and Greatpond Hill.

Greatpond Hill Road (N-S):

0.0 Junction with Route 60 at mile 0.7(0.8). At a three-way fork 15 yards from Route 60, go straight on the middle or E road.

0.5 Main road veers south, changes to dirt, and forks. Take the left dirt road.

0.7 Junction with dirt road. Go straight.

0.85 Two dirt roads to right or W, 10 yards apart. To continue to viewpoints, go left or E.
-The second (southern) road to right or W is the exit to Tipperary Road (0.75 mile).

Greatpond Hill Road

Exit to Tipperary Road:

E-W (W-E)

0.0 (0.7) Junction with Greatpond Hill Road at mile 0.85

0.45 (0.3) Junction. To connect to Tipperary Road, make a left or S turn here. For the view, veer right or NW, and then 10 yards further, veer left. Drive uphill 0.1 mile to house site with **a panoramic view of Great Pond**. The Great Pond Bay's barrier reef is about 0.5 mile offshore, protecting a very large sand-bottomed lagoon. The saltpond berm separates the large saltpond and salt flats from the Bay. Looking north from here one can see the north shore and Green Cay.

0.6 (0.2) Junction with paved Kenjo Drive. Turn right or W onto Kenjo to go to Tipperary Road. On the SW corner of this junction is a **Gregre Tree** with dense, hard wood.

0.75 (0.0) Junction of Kenjo Drive and Tipperary Road, Route 60 [p. 178, mile 1.4(0.1)].

Greatpond Hill Road, N-S (continued):

1.2 Junction as you make a hairpin turn and begin traveling N uphill
-The first two right or southern roads form a 0.55 mile loop that gives **good views of the southern coast and Great Pond area** from the southern peak of Greatpond Hill.

1.45 A left uphill as main road swings W goes 0.05 mile to the top of a Greatpond Hill (640 feet) with **spectacular views of Christiansted and the north shore** from Judith's Fancy to Green Cay and a **high view of the Great Pond area and southern coast** from Robin Bay to Fareham Bay. The northeastern coast view is like that described on p. 175, mile 0.4(1.1), but from a much higher elevation.

1.55 Fork. **Good views of Judith's Fancy, Christiansted, and the Great Pond area** from this fork. Turn left.

1.6 A good view of both the Great Pond area on the south shore and Christiansted on the north.
-From here you must retrace your path to the junction at mile 0.85. From there follow directions to exit on Tipperary Road.

<0.3>

Tipperary Road, Route 60 (continued)

N-S (S-N)

1.0 (0.5) **Mill and ruins of Estate Sight** incorporated into private residence on the right or W

N-S (S-N)
<0.1>
1.1 (0.4) On the right or western side as the road swings E is an experimental planting by the College of the Virgin Islands (now a University), which includes many **Calabash Trees**. These trees have a distinctive growth pattern, and once you learn it, you can recognize them anywhere. They usually have multiple trunks and many spindly branches pointing in graceful arcs out from the center of the tree, much like a fountain. These arcs have few branches, contributing to the fountain-spray look. If the fruits are in season, the green, cannonball-sized gourds will make identification even easier. The **mill of Estate Sight** is on a low hill to the NW.
<0.2>
1.3 (0.2) Dirt road to left or E is residential
<0.1>
1.4 (0.1) Paved road to left or E is Kenjo Drive [p. 177, mile 0.75(0.0)]. This is the connecting road from the south to mile 0.85 of the Greatpond Hill Road. Details of that route are given above on p. 175, mile 0.7(0.8).
<0.1>
1.5 (0.0) Junction with Route 624 [p. 167, mile 6.3(8.0)].

Fig. 58. You may find a surprising toy for a baby on some of your walks. This is a Kallaloo Crab skeleton.

The visit to Buck Island National Monument is often the highlight of a person's trip to St. Croix. It is the largest of the islands offshore from St. Croix and is administered by the National Park Service office in the Christiansted Fort. A free brochure about Buck Island is available in the Fort. Several charter boats, motor or sail, leave the Christiansted wharf daily for half-day and all-day trips. The main reason people choose to visit Buck is to **snorkel the reef** over a marked underwater trail. The water is usually clearer than it is near the shoreline of St. Croix, and the abundance and variety of fish to be seen is enhanced by the prohibition of spearfishing and fish traps. A coral reef practically surrounds Buck, making for excellent **SCUBA diving** (no SCUBA allowed on the snorkel trail). Arrange with a dive boat operator.

While we do not dispute the beauty of the Buck Island reef, we feel that most visitors miss another treasure of the Monument, the island itself. Most of the charter boats stop for lunch, recreation, and restrooms at Turtle Beach on the west end of the island. Few visitors take this opportunity to **hike** a marked, well-maintained trail over the western half of the island. Round trip is 1.4 miles, elevation gain is about 330 feet, and an easy pace takes about 1 hour. From the west-end beach, walk along the south shore until you pass the concrete pier on Diedrich's Point. Turn inland at the south shore picnic area and restrooms; a sign marks the start of the trail. The picnic area is next to a saltpond with many **Black Mangrove Trees** surrounding it. From here it is 400 yards to the summit, Bird Peak at 329 feet; from that point it is all downhill through the interior forest back to the west-end beach. You will see white-flowered **Wild Frangipani, Turk's Cap Cactus** (barrel cactus with red top), **Dildo Cactus** (tall, columnar cactus), **Sandy Point Orchids** growing in the crowns of the dildo, and **Gumbo-limbo Trees** (with reddish peeling bark) thickly laden with epiphytes. The **Caribbean Elaenia** and **Black-whiskered Vireo** can be heard in the forest, but it takes some diligence to see them. Off the main trail a short distance west of the summit is an overlook of the north shore. From the overlook you may be able to view a **Brown Pelican rookery** and **roosting Magnificent Frigatebirds.** A few hundred feet before the trail ends at the west end, it goes through a flat sandy area in a manchineel forest. The green manchineel apples are poisonous. Along the trail in this area are several shrubs with red fleshy fruits one-half inch in diameter, and shiny green leaves. The harmless-looking leaves are lined on the underside with golden needles which readily detach and obnoxiously lodge in the skin of anyone brushing against the plant. This is **Touch-me-not,** known only from St. Croix and Buck Island.

Fig. 59. Trail through the forest on Buck Island.

INDEX TO ORGANISMS WITH COMMON AND SCIENTIFIC NAMES

Acacia sp. 47,60,173
Acropora palmata (Elkhorn Coral) 160
Acrostichum danaeifolium (Swamp Fern) 84,103,104
Adansonia digitata (Baobab Tree, Dead-rat Tree) 35,74,143
Agave, St. Croix (*Agave eggersiana*) 164,171,172
Agave eggersiana (St. Croix Agave) 164,171,172
Albizia lebbeck (Tibet Tree, Old-woman's-tongue, Shek-shek) 71,86
Alexander Palm (Solitaire Palm, *Ptychosperma elegans*) 100
Almond, Indian (*Terminalia catappa*) 34,100
Ameiva polops (St. Croix Ground Lizard) 18,145
Ananas comosus (Pineapple) 54
Anas bahamensis (White-cheeked Pintail) 146,174
Annona muricata (Soursop Tree) 67,102
Annona squamosa (Sugar-apple) 67
Anolis acutus (Tree Lizard) 135
Antennarius spp. (Frogfishes) 65
Anthurium cordifolium (Wild Tanier) 117,118
Ardea herodias (Great Blue Heron) 166
Artibeus jamaicensis (Fruit Bat) 37
Artocarpus altilis (Breadfruit Tree) 54,67,85,87
Australian-pine = Horsetail Tree
Avicennia germinans (Black Mangrove Tree) ....
39,48,49,53,55,56,103,105,140,145,146,149,165,167,174,179
Avocado Tree (*Persea americana*) 54

Banana Tree (*Musa acuminata*, various cultivars) 33,54,102
Banyan Tree (*Ficus* sp.) 37,62
Baobab Tree (Dead-rat Tree, *Adansonia digitata*) 35,74,143
Barracuda, Great (*Sphyraena barracuda*) 140
Barringtonia asiatica 101
Bat, Fruit (*Artibeus jamaicensis*) 37
Batfish (*Ogcocephalus* spp.) 65
Belonidae (Needlefish, Gar) 139
Benjamin Fig (*Ficus benjamina*) 99
Black Calabash (*Enallagma latifolia*) 100
Black Mangrove Tree (*Avicennia germinans*) ....
39,48,49,53,55,56,103,105,140,145,146,149,165,167,174,179
Black Witch (*Erebus odora*) 76
Black-bellied Plover (*Pluvialis squaterola*) 166
Black-crowned Night Heron (*Nycticorax nycticorax*) 174
Black-necked Stilt (*Himantopus mexicanus*) 58,166,174
Black-whiskered Vireo (*Vireo altiloquus*) 166,179
Bread-and-cheese = Catsclaw
Breadfruit Tree (*Artocarpus altilis*) 54,67,85,87
Bridled Quail-Dove (*Geotrygon mystacea*) 77,85,113
bromeliad, large (*Tillandsia utriculata*) 126
Brown Booby (*Sula leucogaster*) 64
Brown Pelican (*Pelecanus occidentalis*) 64,146,166,179
Bubulcus ibis (Cattle Egret) 62,64,170
Bucida buceras (Gregre Tree) 102,103,132,175,177
Bufo marinus (Crapo, Giant Toad) 42

Bursera simaruba (Gumbo Limbo Tree, Naked Tourist Tree, Turpentine Tree) 29,44,47,134,179
Butorides striatus (Green-backed Heron) 85,166

Cacao (pronounced kah-kah'oh, Chocolate, *Theobroma cacao*) 89
Cactus, Dildo (*Pilosocereus royenii*) 156,157,162,179
cactus, Night-blooming Cereus (*Selenicereus urbanianus*) 62
Cactus, Padded Tree (*Opuntia rubescens*) 156
Cactus, Turk's Cap (*Melocactus intortus*) 155,157,179
Cajanus cajan (Pigeon Peas) 54
Calabash Tree (*Crescentia cujete*) 178
Calabash, Black (*Enallagma latifolia*) 100
Calophyllum brasiliense (Santa-maría, False-mamey) 101
Cane, Sugar (*Saccharum officinarum*) 54,102
Cardisoma guanhumi (Kallaloo Crab, Land Crab, White Crab) . 103,105,178
Caribbean Elaenia (*Elaenia martinica*) 60,166,179
Caribbean Martin (*Progne dominicensis*) 64
Caribee Royal Palm (Cuban Royal Palm, South American Royal Palm, *Roystonea oleracea*) 33,100
Carica papaya (Papaya Tree) 54
Casmerodius albus (Great Egret) 166
Cassava (Manioc, *Manihot esculenta*) 54
Cassia fistula (Golden Shower Tree) 20
Cassiopaea frondosa (Upside-down Jellyfish) 140,145
Casuarina equisetifolia (Horsetail Tree, Australian-pine, Horsetail Casuarina) 69
Catsclaw (Bread-and-cheese, *Pithecellobium unguis-cati*) 57
Cattle Egret (*Bubulcus ibis*) 62,64,170
Cedar, Cigar-box (West Indian Cedar, *Cedrela odorata*) 46,80
Cedar, West Indian (Cigar-box Cedar, *Cedrela odorata*) 46,80
Cedrela odorata (Cigar-box Cedar, West Indian Cedar) 46,80
Ceiba (pronounced "say-bah") = Kapok Tree
Cereus, Night-blooming (*Selenicereus urbanianus*) 62
Charadrius semipalmatus (Semi-palmated Plover) 166
Charadrius wilsonia (Wilson's Plover) 166
Chelonia mydas (Green Sea Turtle) 57
Chickrassia tabularis (Chittagong Tree) 46
Chicle = Sapodilla Tree
Chittagong Tree (*Chickrassia tabularis*) 46
Chocolate = Cacao
Christmas Orchid (*Epidendrum ciliare*) 80,124
Christmas Palm = Manila Palm
Cigar-box Cedar (West Indian Cedar, *Cedrela odorata*) 46,80
Cissus sp. (vines) 79,112,135
Citharexylum fruticosum (Pasture Fiddlewood) 69
Clusea rosea (Pitch-apple, False-mamey, Strangler Fig, Wild Fig) 101
Coccoloba uvifera (Seagrape Tree) 60,141,146,150,162,172
Coccothrinax alta (Tyre Palm) 117
Coconut Palm (*Cocos nucifera*) 94,95,97,102,109,110,113,148,156,162
Cocos nucifera (Coconut Palm) 94,95,97,102,109,110,113,148,156,162
Columba leucocephala (White-crowned Pigeon) 60,146,166,174
Columba squamosa (Scaly-naped Pigeon) 150

Common Screwpine (*Pandanus utilis* 100
Common Snipe (*Gallinago gallinago*) 174
Coral, Elkhorn (*Acropora palmata*) 160
Cordia alba (White Manjack) 148,150
Cordia rickseckeri (Fustick, San Bartolomé) 154
Crab, Kallaloo (Land Crab, White Crab, *Cardisoma guanhumi*) 103,105,178
Crapo (Giant Toad, *Bufo marinus*) 42
Crescentia cujete (Calabash Tree) 178
Cuban Royal Palm = Caribee Royal Palm

Dactylopterus volitans (Flying Gurnard) 65
Dalbergia sissoo (Indian Rosewood) 100
Damselfish (family Pomacentridae) 140
Dead-rat Tree (Baobab Tree, *Adansonia digitata*) 35,74,143
Delonix regia (Flamboyant Tree) 68,72,86,99,101
Dendroica petechia (Yellow Warbler) 60,166
Dermochelys coriacea (Leatherback Sea Turtle) 40,59,142
Desert Fan Palm (*Washingtonia fillifera*) 100
Dildo Cactus (*Pilosocereus royenii*) 156,157,162,179
Dwarf Royal Palm = Manila Palm

Echinometra lucunter (Rock-boring Sea Urchin) 58,79,154
Egretta caerulea (Little Blue Heron) 62,64
Elaenia martinica (Caribbean Elaenia) 60,166,179
Elkhorn Coral (*Acropora palmata*) 160
Enallagma latifolia (Black Calabash) 100
Epidendrum bifidum (Chocolate Hole Orchid, Sandy Pt. Orchid) 60,179
Epidendrum ciliare (Christmas Orchid) 80,124
Erebus odora (Black Witch) 76
Eretmochelys imbricata (Hawksbill Turtle) 57
Euphorbia tirucalli (Pencil Tree) 34,88

False-mamey--see Pitch-apple, Santa-maría 101
Fan Palm, Desert (*Washingtonia fillifera*) 100
Fan Palm, Fiji (*Pritchardia pacifica*) 100
Fern, Long-leaved Ruin (*Pteris longifolia*) 96
Fern, Short-leaved Ruin (*Pteris biaurita*) 96
Fern, Strap (*Polypodium phyllitidis*) 75,77,84,117,124
Fern, Swamp (*Acrostichum danaeifolium*) 84,103,104
Ficus benjamina (Benjamin Fig) 99
Ficus citrifolia (Shortleaf Fig, Strangler Fig, White Fig) 62,78,149
Ficus elastica (India Rubber Tree) 28
Ficus obtusifolia (Jagüey, Strangler Fig) 75,76,84,86,134
Ficus sp. (Banyan Tree) 37,62
Fig, Shortleaf (*Ficus citrifolia*) 62,78,149
Fig, Strangler (Shortleaf Fig, White Fig, *Ficus citrifolia*) 62,78,149
Fig, Strangler (Jagüey, *Ficus obtusifolia*) 75,76,84,86,134
Fig, Strangler--also see Pitch-apple
Fig, White = Shortleaf Fig
Fig, Wild = Pitch-apple
Fiji Fan Palm (*Pritchardia pacifica*) 100
Flamboyant Tree (*Delonix regia*) 68,72,86,99,101

Flying Gurnard (*Dactylopterus volitans*) 65
Frangipani, Wild (*Plumeria alba*) 179
Fregata magnificens (Magnificent Frigatebird) 64,166,179
Frogfishes (*Antennarius* spp.) 65
Fruit Bat (*Artibeus jamaicensis*) 37
Fustick (San Bartolomé, *Cordia rickseckeri*) 154

Gallinago gallinago (Common Snipe) 174
Gar (Needlefish, family Belonidae) 139
Garden Eel (*Nystactichthys halis*) 65
Genip Tree (*Melicoccus bijugatus*) 64
Geotrygon mystacea (Bridled Quail-Dove) 77,85,113
Golden Shower Tree (*Cassia fistula*) 20
Grapes (*Vitis* sp.) 54
Grass, Manatee (*Syringodium filiforme*) 147
Grass, Shoal (*Halodule wrightii*) 147
Grass, Turtle (*Thalassia testudinum*) 147
Gray Kingbird (*Tyrannus dominicensis*) 41
Great Blue Heron (*Ardea herodias*) 166
Great Egret (*Casmerodius albus*) 166
Green Sea Turtle (*Chelonia mydas*) 57
Green-backed Heron (*Butorides striatus*) 85,166
Gregre Tree (*Bucida buceras*) 102,103,132,175,177
Guaiacum officinale (Common Lignumvitae Tree) 19,29,45,47,52,64,146,148,173
Gumbo-limbo Tree (Naked Tourist Tree, Turpentine Tree, *Bursera simaruba*) 29,44,47,134,179

Halfbeak (family Hemiramphidae) 139
Halodule wrightii (Shoal Grass) 147
Hawksbill Sea Turtle (*Eretmochelys imbricata*) 57
Hemiramphidae (Halfbeak) 139
Herpestes auropunctatus (mongoose) 18,113
Hibiscus tiliaceus (Beach Mahoe, Sea Hibiscus) 100
Himantopus mexicanus (Black-necked Stilt) 58,166,174
Hippocampus sp. (Seahorses) 65
Hippomane mancinella (Manchineel Tree) 56,56,109,149,150,156,163,163,167,169,179
Hogplum Tree (*Spondias mombin*) 86,89
Honduran Mahogany Tree (*Swietenia macrophylla*) 44,134
Horseradish Tree (*Moringa oleifera*) 54
Horsetail Tree (Australian-pine, Horsetail Casuarina, *Casuarina equisetifolia*) 69
Hura crepitans (Monkey-pistol, Sandbox Tree) 89,90,92
Hymenaea courbaril (West Indian Locust) 112,130
Hymenocallis caribaea (West Indian Spider Lily) 69,123

Iguana iguana (Iguana) 153
India Rubber Tree (*Ficus elastica*) 28
Indian Almond (*Terminalia catappa*) 34,100
Indian Rosewood (*Dalbergia sissoo*) 100
Isognomon sp. (Tree Oyster) 104

Jack Spaniard, larger (*Polisthes crinitus*, illus.) 111
Jack Spaniard, smaller (*Mischocyttarus phthisicus*) 111
Jagüey (*Ficus obtusifolia*, Strangler Fig) 75,76,84,86,134
Jellyfish, Upside-down (*Cassiopaea frondosa*) 140,145
Jojoba (*Simmondsia chinensis*) 54

Kallaloo Crab (Land Crab, White Crab, *Cardisoma guanhumi*) . 103,105,178
Kapok Tree (Ceiba, Silk Cotton Tree, *Ceiba pentandra*) ....
67,69,72,89,89,91,92
Kigelia pinnata (Sausage Tree) 27,53,100

Lablab purpureus (Lablab) 54
Laguncularia racemosa (White Mangrove Tree) ....
39,48,53,55,55,56,63,103,105,140,145,146,146,149,165,167,169,174,179
Larus atricilla (Laughing Gull) 64
Laughing Gull (*Larus atricilla*) 64
Least Tern (*Sterna antillarum*) 58,60,166
Leatherback Sea Turtle (*Dermochelys coriacea*) 40,59,142
Leucaena leucocephala (Tantan Tree) 47,54
Lignumvitae Tree (*Guaiacum officinale*) 19,29,45,47,52,64,146,148,173
Lily, West Indian Spider (*Hymenocallis caribaea*) 69,123
Little Blue Heron (*Egretta caerulea*) 62,64
lizard, Iguana (*Iguana iguana*) 153
Lizard, St. Croix Ground (*Ameiva polops*) 18,145
Lizard, Tree (*Anolis acutus*) 135
Loblolly = Water Mampoo
Locust, West Indian (*Hymenaea courbaril*) 112,130
Lutjanidae (Snappers) 139
Lutjanus apodus (Schoolmaster) 140

Magnificent Frigatebird (*Fregata magnificens*) 64,166,179
Mahoe, Beach (Sea Hibiscus, *Hibiscus tiliaceus*) 100
Mahogany Tree, Honduran (*Swietenia macrophylla*) 44,134
Mahogany Tree, hybrid (*Swietenia* hybrid) 29,44,45,49
Mahogany Tree, Pacific (*Swietenia humilis*) 43,44,45
Mahogany Tree, West Indian (*Swietenia mahagoni*) ....
19,21,44,45,66,67,68,69,81,102,109,123,135
Malpighia infestissima (Touch-me-not) 149,179
Mammea americana (Mammee Apple) 86,89,112
Mammee Apple (*Mammea americana*) 86,89,112
Mampoo Tree, Water (Loblolly, *Pisonia subcordata*) 161
Manatee Grass (*Syringodium filiforme*) 147
Manchineel Tree (*Hippomane mancinella*) ....
56,56,109,149,150,156,163,163,167,169,179
Mangifera indica (Mango Tree) 33,54,67,83,86,88,91
Mango Tree (*Mangifera indica*) 33,54,67,83,86,88,91
Mangrove Tree, Black (*Avicennia germinans*) ....
39,48,49,53,55,56,103,105,140,145,146,149,165,167,174,179
Mangrove Tree, Red (*Rhizophora mangle*) ....
39,48,49,53,55,56,103,104,105,145,146,149,165,167,174
Mangrove Tree, White (*Laguncularia racemosa*) ....
39,48,53,55,56,63,103,105,140,145,146,149,165,167,169,174,179

Manihot esculenta (Cassava, Manioc) 54
Manila Palm (Christmas Palm, Dwarf Royal Palm, *Veitchia merrillii*) ... 100
Manilkara zapota (Chicle, Mespel Tree, Sapodilla Tree) 54,58
Manioc (Cassava, *Manihot esculenta*) 54
Manjack, White (*Cordia alba*) 148,150
Margarops fuscatus (Pearly-eyed Thrasher) 77,91
Melicoccus bijugatus (Genip Tree) 64
Melocactus intortus (Turk's Cap Cactus) 155,157,179
millipede, black (*Orthocricus arboreus*) 77
millipede, brick red (*Trigoniulus lumbricinus*) 77
millipede, brown with yellow-green bars (*Orthocricus monilicornis*) 77
Mischocyttarus phthisicus (smaller Jack Spaniard) 111
Molly (*Poecilia* sp.) 83,85
mongoose (*Herpestes auropunctatus*) 18,113
Monkey-pistol Tree (Sandbox Tree, *Hura crepitans*) 89,90,92
Morinda citrifolia (Painkiller Tree) 54
Moringa oleifera (Horseradish Tree) 54
Mother-in-law's Tongue = Snake Plant
Mugil sp. (Mullet, family Mugilidae) 139,145
Mullet (family Mugilidae) 139,145
Musa acuminata, various cultivars (Banana Tree) 33,54,102

Nasutitermes costalis (Long-nosed Termite) 15
Needlefish (family Belonidae) 139
Nephila clavipes (Golden Silk Orb Weaver Spider) 76,85,118,119
Night-blooming Cereus (*Selenicereus urbanianus*) 62
Nycticorax nycticorax (Black-crowned Night Heron) 174
Nycticorax violaceus (Yellow-crowned Night Heron) 85
Nystactichthys halis (Garden Eel) 65

Ogcocephalus spp. (Batfish) 65
Old Man's Beard (*Tillandsia recurvata*) 19
Old-woman's-tongue Tree = Tibet Tree
Opuntia rubescens (Padded Tree Cactus) 156
Orchid, Chocolate Hole (Sandy Pt. Orchid, *Epidendrum bifidum*) 60,179
Orchid, Christmas (*Epidendrum ciliare*) 80,124
Orchid, Sandy Pt. (Chocolate Hole Orchid, *Epidendrum bifidum*) 60,179
Orchid, Vanilla (*Vanilla planifolia*) 125
Orthocricus arboreus (black millipede) 77
Orthocricus monilicornis (millipede, brown with yellow-green bars) 77
Otaheita (pronounced "o-tah-high'tah", *Thespesia populnea*) 109,150
Oyster, Tree (*Isognomon* sp.) 104

Pacific Mahogany Tree (*Swietenia humilis*) 43,44,45
Painkiller Tree (*Morinda citrifolia*) 54
Palm, Alexander (Solitaire Palm, *Ptychosperma elegans*) 100
Palm, Caribee Royal (Cuban Royal Palm, South American Royal Palm, *Roystonea oleracea*) 33,83,100
Palm, Coconut (*Cocos nucifera*) 94,95,97,102,109,110,113,148,156,162
Palm, Cuban Royal = Caribee Royal Palm
Palm, Desert Fan (*Washingtonia fillifera*) 100

Palm, Dwarf Royal = Manila Palm
Palm, Manila (Christmas Palm, Dwarf Royal Palm, *Veitchia merrillii*) ...100
Palm, Puerto Rican Royal (*Roystonea borinquena*) 69,83,86,100
Palm, Solitaire = Alexander Palm
Palm, South American Royal = Caribee Royal Palm
Palm, Tyre (*Coccothrinax alta*) 117
Palmetto, Puerto Rican (*Sabal causiarum*) 100
Pandanus utilis (Common Screwpine) 100
Papaya Tree (*Carica papaya*) 54
Passiflora edulis (passion fruit) 54
passion fruit (*Passiflora edulis*) 54
Pasture Fiddlewood (*Citharexylum fruticosum*) 69
Pearly-eyed Thrasher (*Margarops fuscatus*) 77,91
Pelecanus occidentalis (Brown Pelican) 64,146,166,179
Peltophorum pterocarpum (Yellow Flamboyant) 64
Pencil Tree (*Euphorbia tirucalli*) 34,88
Persea americana (Avocado Tree) 54
Phaethon lepturus (White-tailed Tropicbird) 53,171
Pigeon Peas (*Cajanus cajan*) 54
Pilosocereus royenii (Dildo Cactus) 156,157,162,179
Pineapple (*Ananas comosus*) 54
Pink-cedar Tree = White-cedar Tree
Pisonia subcordata (Water Mampoo Tree, Loblolly) 161
Pitch-apple (False-mamey, Strangler Fig, Wild Fig, *Clusea rosea*) 101
Pithecellobium unguis-cati (Catsclaw, Bread-and-cheese) 57
Plumeria alba (Wild Frangipani) 179
Pluvialis squaterola (Black-bellied Plover) 166
Poecilia sp. (Molly) 83,85
Polisthes crinitus (larger Jack Spaniard) 111
Polypodium phyllitidis (Strap Fern) 75,77,84,117,124
Pomacentridae (Damselfish) 140
Primavera (*Roseodendron donnell-smithii*) 80
Pritchardia pacifica (Fiji Fan Palm) 100
Progne dominicensis (Caribbean Martin) 64
Pteris biaurita (Short-leaved Ruin Fern) 96
Pteris longifolia (Long-leaved Ruin Fern) 96
Ptychosperma elegans (Alexander Palm, Solitaire Palm) 100
Puerto Rican Palmetto (*Sabal causiarum*) 100
Puerto Rican Royal Palm (*Roystonea borinquena*) 69,83,86,100

Raintree (Samán [pronounced say'man], *Samanea saman*) ....
71,72,79,82,86,92,135
Red Mangrove Tree (*Rhizophora mangle*) ....
39,48,49,53,55,56,103,104,105,140,145,146,149,165,167,174
Rhizophora mangle (Red Mangrove Tree) ....
39,48,49,53,55,56,103,104,105,140,145,146,149,165,167,174
Rock-boring Sea Urchin (*Echinometra lucunter*) 58,79,154
Roseodendron donnell-smithii (Primavera) 80
Rosewood, Indian (*Dalbergia sissoo*) 100
Royal Palm, Dwarf = Manila Palm
Royal Palm, Puerto Rican (*Roystonea borinquena*) 69,83,86,100
Roystonea borinquena (Puerto Rican Royal Palm) 69,83,86,100

Roystonea oleracea (Caribbee Royal Palm, Cuban Royal Palm, South American Royal Palm) 33,100

Sabal causiarum (Puerto Rican Palmetto) 100
Saccharum officinarum (Sugar Cane) 54,102
St. Croix Agave (*Agave eggersiana*) 164,171,172
St. Croix Ground Lizard (*Ameiva polops*) 18,145
Samanea saman (Raintree, Samán) 71,72,79,82,86,92,135
Samán = Raintree
San Bartolomé (Fustick, *Cordia rickseckeri*) 154
Sandbox Tree (Monkey-pistol, *Hura crepitans*) 89,90,92
Sandy Pt. Orchid (Chocolate Hole Orchid, *Epidendrum bifidum*) 60,179
Sansevieria trifasciata (Snake Plant, Mother-in-law's Tongue) 34
Santa-maría (False-mamey, *Calophyllum brasiliense*) 101
Sapodilla (Chicle, Mespel Tree, *Manilkara zapota*) 54,58
Sarotherodon mossambica (Tilapia) 54
Sausage Tree (*Kigelia pinnata*) 27,53,100
Scaly-naped Pigeon (*Columba squamosa*) 150
Schoolmaster (*Lutjanus apodus*) 140
Screwpine, Common (*Pandanus utilis*) 100
Sea Hibiscus = Beach Mahoe
Seagrape Tree (*Coccoloba uvifera*) 60,141,146,150,162,172
Seahorses (*Hippocampus* sp.) 65
Selenicereus urbanianus (Night-blooming Cereus) 62
Semi-palmated Plover (*Charadrius semipalmatus*) 166
Shek-shek = Tibet Tree
Shoal Grass (*Halodule wrightii*) 147
Shortleaf Fig (White Fig, Strangler Fig, *Ficus citrifolia*) 62,78,149
Silk Cotton Tree (Ceiba, Kapok Tree, *Ceiba pentandra*) 67,69,72,89,89,91,92
Silver Tabebuia Tree (*Tabebuia argentea*) 99,134
Simmondsia chinensis (Jojoba) 54
Snake Plant (Mother-in-law's Tongue, *Sansevieria trifasciata*) 34
Snappers (family Lutjanidae) 139
Solitaire Palm = Alexander Palm
Sorghum bicolor (Sorghum) 54
Soursop Tree (*Annona muricata*) 67,102
South American Royal Palm = Caribee Royal Palm
Spanish Moss (*Tillandsia usneoides*) 22,90,91
Sphyraena barracuda (Great Barracuda) 140
Sphyrapicus varius (Yellow-bellied Sapsucker) 103,132
Spider Lily, West Indian (*Hymenocallis caribaea*) 69,123
Spider, Golden Silk Orb Weaver (*Nephila clavipes*) 76,85,118,119
Spirula, Common (*Spirula spirula*) 156
Spondias mombin (Hogplum Tree) 86,89
Sterna antillarum (Least Tern) 58,60,166
Strangler Fig (Jagüey, *Ficus obtusifolia*) 75,76,84,86,134
Strangler Fig (Shortleaf Fig, White Fig, *Ficus citrifolia*) 62,78,149
Strangler Fig--also see Pitch-apple
Sugar Cane (*Saccharum officinarum*) 54,102
Sugar-apple (*Annona squamosa*) 67
Sula leucogaster (Brown Booby) 64

Sweet-lime (*Triphasia trifolia*) 46,84
Swietenia humilis (Pacific Mahogany Tree) 43,44,45
Swietenia hybrid (hybrid Mahogany Tree) 29,44,45,49
Swietenia macrophylla (Honduran Mahogany Tree) 44,134
Swietenia mahagoni (West Indian Mahogany Tree) ....
19,21,44,45,66,67,68,69,81,102,109,123,135
Syringodium filiforme (Manatee Grass) 147

Tabebuia argentea (Silver Tabebuia Tree) 99,134
Tabebuia heterophylla (White-cedar Tree, Pink-cedar Tree) 50,53,86,134
Tabebuia, Silver Tabebuia Tree (*Tabebuia argentea*) 99,134
Tamarind Tree (*Tamarindus indica*) 68,69,89,167
Tamarindus indica (Tamarind Tree) 68,69,89,167
Tantan Tree (*Leucaena leucocephala*) 47,54
Teak Tree (*Tectona grandis*) 6,42,44,97,120
Tectona grandis (Teak Tree) 6,42,44,97,120
Terminalia catappa (Indian Almond) 34,100
Termite, Long-nosed (*Nasutitermes costalis*) 15
Thalassia testudinum (Turtle Grass) 147
Theobroma cacao (Cacao [pronounced kah-kah'oh], Chocolate) 89
Thespesia populnea (Otaheita) 109,150
Tibet Tree (Old-woman's-tongue, Shek-shek, *Albizia lebbeck*) 71,86
Tilapia (*Sarotherodon mossambica*) 54
Tillandsia recurvata (Old Man's Beard) 19
Tillandsia usneoides (Spanish Moss) 22,90,91
Tillandsia utriculata (large bromeliad) 126
Toad, Giant (Crapo, *Bufo marinus*) 42
Touch-me-not (*Malpighia infestissima*) 149,179
Tourist Tree, Naked = Gumbo-limbo Tree
Trigoniulus lumbricinus (brick-red millipede, gungalo) 77
Triphasia trifolia (Sweet-lime) 46,84
Turk's Cap Cactus (*Melocactus intortus*) 155,157,179
Turpentine Tree = Gumbo-limbo Tree
Turtle Grass (*Thalassia testudinum*) 147
Turtle, Green Sea (*Chelonia mydas*) 57
Turtle, Hawksbill (*Eretmochelys imbricata*) 57
Turtle, Leatherback Sea (*Dermochelys coriacea*) 40,59,142
Tyrannus dominicensis (Gray Kingbird) 41
Tyre Palm (*Coccothrinax alta*) 117

Upside-down Jellyfish (*Cassiopaea frondosa*) 140,145

Vanilla planifolia (Vanilla Orchid) 125
Veitchia merrillii (Manila Palm, Christmas Palm, Dwarf Royal Palm) ... 100
vine, *Cissus* sp. 79,112,135
Vireo altiloquus (Black-whiskered Vireo) 166,179
Vitis sp. (grapes) 54

Washingtonia fillifera (Desert Fan Palm) 100
Water Mampoo Tree (Loblolly, *Pisonia subcordata*) 161
West Indian Cedar (Cigar-box Cedar, *Cedrela odorata*) 46,80
West Indian Locust (*Hymenaea courbaril*) 112,130

West Indian Mahogany Tree (*Swietenia mahagoni*) ....
19,21,44,45,66,67,68,69,81,102,109,123,135
West Indian Spider Lily (*Hymenocallis caribaea*) 69
White Fig = Shortleaf Fig
White Mangrove Tree (*Laguncularia racemosa*) ....
39,48,53,55,56,63,103,105,140,145,146,149,165,167,169,174,179
White Manjack (*Cordia alba*) 148,150
White-cedar Tree (Pink-cedar Tree, *Tabebuia heterophylla*) 50,53,86,134
White-cheeked Pintail (*Anas bahamensis*) 146,174
White-crowned Pigeon (*Columba leucocephala*) 60,146,166,174
White-tailed Tropicbird (*Phaethon lepturus*) 53,171
Wild Fig = Pitch-apple
Wild Frangipani (*Plumeria alba*) 179
Wild Tanier (pronounced "tan-yah", *Anthurium cordifolium*) 117,118
Wilson's Plover (*Charadrius wilsonia*) 166

Yellow Flamboyant (*Peltophorum pterocarpum*) 64
Yellow Warbler (*Dendroica petechia*) 60,166
Yellow-bellied Sapsucker (*Sphyrapicus varius*) 103,132
Yellow-crowned Night Heron (*Nycticorax violaceus*) 85

GENERAL INDEX

i = illustration

Acacia 47,60,173
Agave, St. Croix 164,171,172i
Agricultural and Food Fair 33
Agricultural Station, federal and University of the Virgin Islands 32,54
  Virgin Islands Government 33
Airport Road (Route 64) 54-57
airport, international 56
  seaplane 21;24 view
algal ridge, Isaac Point 157, 158i
  Robin Bay 164
All-for-the-Better Road 175
Almond, Indian 34,100
Altona Lagoon 139-140,144; views 24,138,141,142,143
Altona Lagoon inlet 139
Altona Lagoon Salt Flats 144
Alucroix Channel 48-49; views 23,139
Annaly Bay 119,122,123; area map 117
Annaly Bay Notch 119,122
Annaly Road 88
Anthurium 117,118
Apothecary Hall, Frederiksted 69
aqueduct, LaGrange 81
Arawak Indians 106
Archaelogy, Department of 69
artifacts 17,64,65,69
Atlantic Fleet Weapons Training Facility 38,73
Australian-pine = Horsetail Tree
Avocado Tree 54

baby house, Estate Betty's Hope 57
  Estate Canegarden 173
  Estate Castle Coakley 26
ballcourt, pre-Columbian Indian 106
banana grove 33,54,102
Banyan Tree 37,62
Baobab Tree (Dead-rat Tree) 35,74,143
Barracuda, Great 140
barrier rampart 169
barrier reef, Bluegut Bay 155
  Boiler Bay 155
  Tague Bay 151
Barringtonia 101
basketry, Sheltered Workshop 29
Bassin Triangle 21
Bat, Fruit 37
Batfish 65
Baths, Malta (Monks) 78
Batist Point (Fishermen's Point) 172
Battery, Danish 148
beach, Altona Lagoon 140
  Baron Bluff 107
  Beauregard Beach and Bay 141,144i
  Boiler Bay (Brown Bay, Whale Bay) 155
  Buck Island 179
  Butler Bay 74,75
  Cane Bay 110,112
  Chenay Bay 147
  Coakley Bay 148
  Columbus' Landing 104
  Cottongarden Bay 153
  Cramer Park 153
  Davis Beach 113
  East End Bay 156
  Fairleigh Dickinson Territorial Park 153-157
  Fort Louise Augusta 141
  Frederiksted public 69
  Grapetree Bay 160, (Meggie) 161
  Grapetree Beach Hotel 162
  Great Pond Bay 164,169
  Hodge Bay 150
  Isaac Bay 157
  Jack Bay 157
  Judith's Fancy 97
  LaGrange 69,72
  LaVallee 110
  Little Princess 95
  Manchenil Bay 171
  Meggie (Grapetree Bay) 161
  North Star 113
  Northside 78
  Pelican Cove 97
  Prune Bay 147,148
  Punnett Bay 142
  Rainbow Beach 72
  Reef Condominium 151
  Romney Point 152
  Rust-op-twist 108,109
  Sandy Point 59,63
  Shoy's Beach 142
  Smuggler's Cove 152
  Solitude Bay 150
  Southgate 146
  Sprat Hall 72
  Stony Ground 63
  Tamarind Reef 146
  Turner Hole 162

beach, Turtle Beach, Buck Island 179
  Whistle Beach 144
  Yellowcliff Bay 151
beachrock, Annaly area 118,119,123
  Ham Bluff 80
  Malta Baths 79
  LaVallee 110
  Whiteladies 58
  Wills Bay 118
Beauregard Beach and Bay 141,144i
Bee Peak 162
Benjamin Fig 99
Benjamin House, Frederiksted 69
Bennykenny Hill 114,133
Beresford Manor 78
Bethlehem Gut 87
Bethlehem Old Works, area map 31
  chimney, view 87
  sugar factory 30,31
Bird Peak 179
birdwatching, Altona Lagoon 140,144
  Alucroix Channel 48
  Buccaneer Hotel 144
  Canegarden Bay 53,173
  Carambola Golf Course 134
  Castle Burk pond 87
  Coakley Bay Saltpond 148,149
  Fredensborg Pond 30
  Frederiksted 64
  Frederiksted pier 64
  French Pond 53,173
  Great Pond 165,166
  Krause Lagoon 55
  Long Point mangrove 39
  Manchenil Bay 171
  Manning Bay Lagoon 55,56
  Mt. Fancy Saltpond 164
  Prosperity Marsh 72
  Rust-op-twist Marsh 109
  Sandy Point 40,60
  Southgate Saltpond 146,147
  Westend Saltpond 40,58,62
Black Calabash 100
black coral, Cane Bay 110
  Frederiksted reef 65
Black Witch 76
Black-bellied Plover 166
Black-crowned Night Heron 174
Black-necked Stilt 58,166,174
Black-whiskered Vireo 166,179
blacksmith, St. George Botanical Gardens, 36i
Blue Mountain 33,88,132
Bluegut Bay (François Bay) 155
Bluegut Bay (Lamb Bay) 154i,155
Boiler Bay (Brown Bay, Whale Bay) 153,155
Botanical Garden, St. George 35
botanical garden--also see horticultural site
Boy Scout Camp, Great Pond 169
Bread-and-cheese = Catsclaw
Breadfruit 67i
Breadfruit Tree 54,85,87
Bridled Quail-Dove 77,85,113
bromeliad, large 126i
Brown Bay (Boiler Bay, Whale Bay) 155
Brown Booby 64
Brown Pelican 64,146,166
  rookery 179
Buccaneer Estates 142
Buccaneer Hotel area 142-145
Buck Island National Monument 18,179,180i
  for cruise ship passengers 14
  tours 17,143,151
  views 150,151,152,168
Butler Bay Gut 75-78

Cacao 89
Cactus, Dildo 156i,157,162,179
  Night-blooming Cereus 62
  padded tree 156i
  Turk's Cap 155i,157,179
Calabash Tree 178
Calabash, Black 100
Caledonia formation 123,124,158
Caledonia Gut 79,123,125i
Caledonia Mountain 116
Caledonia Peak 108
Caledonia Valley 124
Camp Arawak 164
campground and camping,
  Camp Arawak 164
  Cramer Park 153
  Great Pond 169
Camphouse Cove 141,144
Canaan Ridge Road (Route 73) 129,131-133
Candle Reef (Dry Ledge) 149
Candle Reef, view 24,138
Cane Bay 110-112
Cane Valley Formation 85
cane, sugar 54,102

Canegarden Bay 53,173
Carambola Beach Resort 113
Carambola Golf Course 134
Carden Bay 148,149
Carib Indians 106
Caribbean Elaenia 60,166,179
Caribbean Martin 64
Carina Mountain 129
cars, rental 8
Carty Bay 163
Carty Point 163
Cassava (Manioc) 54
Castle Nugent Farms 169
Catsclaw 57
Cattle Egret 62,64,170i
cattle, Senepol 170i
Cedar, Cigar-box (West Indian Cedar) 46,80
Ceiba = Kapok, Silk Cotton Tree
cemetery, Christiansted 21
  Frederiksted 68
  Holy Trinity Lutheran 67
  St. Patrick's Church 67
Centerline Road (Route 70) 24-41
Centerline Tour 17-41
  for cruise ship passengers 13
Cereus, Night-blooming 62
Channel Rock 160
Chapel of Our Mother of Perpetual Help 84
Chenay Bay 147,148
Chicle = Sapodilla Tree 54
chimney--see Estate
chimney 33
Chittagong Tree 46
Chocolate = Cacao
Christening Hole 80
Christiansted, map 16
  walking tour 18
Christiansted Harbor 17-18,137; views 22i,23,141,143,168
Christiansted National Historic Site, headquarters 17
Christiansted Tour 17-21
Cigar-box Cedar (West Indian Cedar) 46 description; 80
Cissus vine 79,112,135
Cistern Bay (Knight Bay) 152
Clairmont Peak 128
Clindinen Hill 151
clinic, Frederiksted 64
clock tower, Frederiksted 66
Coakley Bay Saltpond 148
Coast Guard Station 80
Coconut Palm 94i,95,97,102,109,110, 113,148,162
Coconut seed 156
Colorado River 155
Columbus, Christopher 9,105
Columbus' Landing 104
Common Snipe 174
Company Street 17,18-19
Concordia Ridge Road 127
Container Port, St. Croix 49
Contentment Road 21
copper 73i
coral 131
Coral, Elkhorn 160
corals, hard, Bluegut Bay 155
  Boiler Bay 155
  Cane Bay 110
  Davis Beach 113
  East End Bay 156
  Frederiksted 65
  Isaac Bay 157
  Malta Baths 79
  (near) Manchenil Bay 172
  North Star 112
  Pelican Cove 97
  Robin Bay 164
  Rod Bay 163
corals, soft, Coakley Bay 149
  Malta Baths 79
Cottongarden Bay 153
Cottongarden Point 153
Cottongrove Bay 164
Cottongrove Strand 164
court, Virgin Islands Territorial 30
crab trap 105i
Crab, Kallaloo (Land Crab, White Crab) 103,105,178i
crafts, basketry (Sheltered Workshop) 29
crafts, ironwork 36
crafts, wood (LEAP) 82
Cramer Park 153
Crapo (Giant Toad) 42
Creque Dam 89,90i
Creque Dam Road (Route 58) 77,88-92,126i
Cross Island Road 175
Cruzan Rum Distillery 57
Customs 17
Customs House, Christiansted 17

Customs House, Frederiksted 66
Cutlass Cove 143

damselfish 140
Danish West India & Guinea Company 10,17
Darby Hill 142
Davis Beach 113
Davis Beach road 113,133
de Poincy 9
Diedrich's Point 179
diente del perro, Annaly area 118,119,123
  Ham Bluff 80
  LaVallee 110
  Malta Baths 79
  Whiteladies 58
  Wills Bay 118
Dildo Cactus 156i,157,162,179
directions 7
distillery, Brugal Rum 72
  Cruzan Rum 57
  tours 57
dive, Buck Island 179
  Cane Bay 110
  Davis Beach 113
  Frederiksted Pier 65
  Grass Point 163
  Ham Bluff 80
  Malta Baths 79
  North Star 112
diving tours 17,64,143,151
diving tours, northwest coast 109,131
Dolby Hill Road 99
drift seeds 156
Dry Ledge (Candle Reef) 149
Duggan's Reef Restaurant 151
Dyck Marine Center 98

Eagle Ridge 33,129
East End and South Shore Tour, for cruise ship passengers 14
East End Bay (Windward Bay) 156
East End Road (Route 82) 137-158
East Point (Kid Peak, Point Udall) 157
Easthill School 148
egrets 146
emancipation 66
epiphyte 91
Estate Adventure (mill) 34,135
  Allandale = Estate Bog of Allan
  Anguilla (mill) 55
Estate Anna's Hope (in Eastend A Quarter) 142
  Anna's Hope (in Company Quarter, also Nancy's Hope, bell tower, mill, village) 25
  Annaly (factory, greathouse, mill) 82,84
  Barren Spot (mill) 27
  Beeston Hill (mill) 24
  Belvedere (in Northside B Quarter, mill) 131
  Belvedere (in Queen Quarter, combined with Estate Rattan, watermill) 96
  Bethlehem Middle Works (mill remnants) 55
  Bethlehem New Works, (village) 55
  Bethlehem Old Works (chimney, sugar factory 31; greathouse, manager's house, village 54,55)
  Betty's Hope (mill) 56
  Big Fountain (also Fountain, factory, mill) 134; view 115
  Bodkin (mill) 12i,119,121i
  Body Slob = Estate Slob (in King Quarter)
  Boetzberg (also Butzberg, mill) 142,169
  Bog of Allan (also Allandale, greathouse) 37
  Bonne Esperance (cane-grinding mill, watermill) 28
  Butler Bay (also called Prospect Hill, factory, greathouse, lime kiln, mill) 74,75
  Butzberg = Estate Boetzberg
  Camporico (also West End, mill) 40
  Cane Bay (animal mill, factory, greathouse, village) 110
  Canegarden (animal mill, baby house, greathouse, mill) 173
  Carden (included in Estate Coakley Bay)
  Carlton (mill) 39
  Castle Burk 32,87
  Castle Coakley (baby house, mills, village) 26,27
  Clairmont (in Northside B Quarter, also Clearmount, mill, sugar factory, village) 108
Estate Clairmont Park 108

Estate Clairmont (in Prince Quarter, also called Clear Mount and Clermont, mill) 37
Estate Clairmont Road 107
Estate Clear Mount = Estate Clairmont in Prince Quarter, also called Clermont
Clearmount = Estate Clairmont in Northside B Quarter
Clermont = Estate Clairmont in Prince Quarter, also called Clear Mount
Clifton Hill (greathouse, cane-grinding mills, watermill) 28,29,47
Coakley Bay (includes Estate Carden, animal mill, factory, greathouse, cane-grinding windmill, village) 148
Coble (village) 87
Colquehoun (also Colquehoun Mount Pleasant, Mount Pleasant in King Quarter, bell tower, cane-grinding mill remnants, watermill) 87
Colquehoun Mount Pleasant = Estate Colquehoun
Concordia (in Queens Quarter, also Upper Concordia, chimney, mill, village) 102
Concordia (in Westend Quarter, also Lower Concordia, South Concordia, mill) 71
Constitution Hill (mill) 43
Cotton Grove (mill) 164
Cotton Valley (mill) 150
Diamond (in Prince Quarter, mill) 57
Diamond (in Queen Quarter, chimney, mill) 27
Diamond Keturah (also Little Diamond, mill) views 23,138,171,173
Die Liefde = Estate Lower Love
Dolby Hill (also Montpellier in Queen Quarter, mill) 99
Enfield Green (greathouse, mills) 51,57
Fareham (mill) 169
Fountain = Estate Big Fountain
Fox (included in Estate Negro Bay)
Fredensborg (mill) 31,132
Estate French Pond 53,173
Glynn (mills) 88,99
Golden Grove (greathouse, mill) 33,51
Golden Rock (mill) 93
Granard (greathouse) 173
Grange (greathouse) 25
Grapetree Bay 160
Great Pond (greathouse) 164
Greencay (chimney, factory, mills) 147
Grove Place (chimney, greathouse) 35
Ham Bay (mill) 79,124
Hannah's Rest (greathouse) 40
Hermitage (factory, mill) 87
Hermon Hill (mill) 22
Hogensborg (also Sorgenfrie, chimney, greathouse) 37i
Hope (mill) 52
Jealousy (mills) 87
John's Rest = Estate Whim
Judith's Fancy (mill) 97,128; views 23,138
King (also Upper Bethlehem, village) 29,31,133
Kingshill (barracks) 29
LaGrande Princess (greathouse, mill) 96
LaGrange (greathouse, mill) 72
LaVallee (chimney, mill) 109,110
Libanon (chimney, cobble road) 102,132
Little Diamond = Estate Diamond Keturah
Little Fountain (mill) 132
Little LaGrange (mill, view) 82
Longford (chimney) 170
Lower Concordia = Estate Concordia of Westend Quarter, also South Concordia
Lower Love (also Die Liefde, New Love, chimney, cobble road, sugar factory) 33; 86 view
Mary's Fancy (chimney) 28
MonBijou (greathouse, mill) 132
Montpellier = Estate Dolby Hill in Queen Quarter
Montpellier (in Prince Quarter, chimney) 84
Morningstar (chimney, mill, view) 129

Estate Mount Pleasant (in King Quarter) = Estate Colquehoun, Estate Colquehoun Mount Pleasant
Mount Pleasant (in Northside A Quarter) = Mount Victory
Mount Pleasant (in Prince Quarter, greathouse) 35
Mount Stewart (includes Estate Wills Bay)
Mount Victory (in Northside A Quarter, also Mount Pleasant, factory, greathouse, mill) 91,92; view 124
Mount Washington (animal mill, cemetery, sugar factory) 74
Mount Welcome (mill) 24 view; 139
Nancy's Hope = Anna's Hope in Company Quarter
Negro Bay (includes Estate Fox, mill) 34
New Love = Lower Love
Nicholas (factory, mill) 92; view 124
North Side (chimney, factory, mills) 78
North Star (mill) 112
Nugent (greathouse) 169
Old Love = Upper Love
Old Pye (mill) 57
Orange Grove (in Company Quarter, greathouse, mill) 21
Orangegrove (in Prince and Westend Quarters, chimney, greathouse, mill, village) 83; view 122
Oxford (greathouse) 82
Paradise (mill) 34,51
Parasol (mill) 134; view 120
Peter's Farm, hospital 21
Peter's Rest (bell tower, chimney, mill) 25,26
Petronella (mill) view 168;169
Plessen (mill) 35
Profit (mill) 29,50
Prospect Hill (called Estate Butler Bay)
Prosperity (in Northside B Quarter, chimney, factory, greathouse, mill) 113
Prosperity (in Westend Quarter, chimney, greathouse, mill remnants, sugar factory, village) 72,81
Estate Prosperity Garden (greathouse) 81
Punch (mill) views 72,82,91
Rattan and Belvedere (in Queen Quarter, chimney, mill) 96,98
Richmond (greathouse) 23,138
River (mill, village) 135; view 115
Rust-op-twist (animal mill, factory, cane-grinding mill, village) 109
St. George (sugar factory, village) 35; watermill 36
St. John (chimney, factory, mill) 97,98
Salt River (mill, view) 105
Shoy (also Shoy's, chimney, mill) 142,143
Sight (mill) view 168;177
Sion Farm (factory, greathouse, mills) 44,96
Sion Hill (chimney, greathouse, mills) 43,44,96
Slob (in King Quarter, also Body Slob, mill, village, watermill) 28,31,133
Smithfield (mill) 41
Sorgenfrie = Estate Hogensborg
South Concordia = Estate Concordia of Westend Quarter, also Lower Concordia
Southgate (greathouse, mills) 146
Spanish Town (greathouse) 49,50i
Sprat Hall (factory, greathouse [hotel], mill) 5i,73,89
Springfield (mill) 84; view 123
Stony Ground 41
Strawberry Hill (mills) 27
Thomas 44
Two Brothers (mill, manager's house) 62
Two Friends (chimney) 84,122
Upper Bethlehem includes Estate King
Upper Concordia = Concordia in Queens Quarter
Upper Love (also Old Love, chimney) 86
West End = Estate Camporico
Whim (also John's Rest, restored animal mill, restored greathouse, museum, sugar

factory, watermill, restored cane-grinding mill) 39,40i
Estate White's Bay (greathouse) 58
William (mill) 72
Wills Bay (included in Estate Mt. Stewart, mill) 119,122,123
Windsor (mill) 102
Evans, Melvin 46
Experimental Forest 25,26,43-46,96; back road 93-95
Fairleigh Dickinson Territorial Park 153-157
Fairleigh Dickinson University 98,152
Fairplain Creek 55
False-mamey--see Pitch-apple, Santa-maría
Fan Palm, Desert 100
Fiji 100
Fanny's Fancy Point 164
Fareham Bay 172
farms, Estate Glynn 102
Fern, Long-leaved Ruin 96
Short-leaved Ruin 96
Strap 75,77,84,117,124
Swamp 84,103i,104
Fig, Benjamin 99
Shortleaf 62,78,149
strangler 75,76i,84,86,134
strangler (Shortleaf) 62,78,149
strangler--also see Pitch-apple
Wild = Pitch-apple
fish market, Christiansted 137
Frederiksted 62,63i
fish trap 70i
Fishermen's Point (Batist Point) 172
fishing tours 17,64,143,151
fishing, bait fish 64,139
fishing, Frederiksted Pier 64
fishing, surf 60
Flamboyant Tree 68,72,86,99,101
Flying Gurnard 65
Forest Tour, for cruise ship passengers 13
forest, Creque Dam Road 72,88-92
Mahogany Road 72,81-85
Fort Christiansvaern 17,18i,137, view 141
Fort Frederik 66
Fort Louise Augusta view 24;141
Fort Salé 106
Fountain Valley Road 133
Four-lane Highway (Route 66) 46-51
Frangipani, Wild 179
François Bay (Bluegut Bay) 155
Fredensborg, area map 31
Fredensborg Pond 30,32i,101
Fredensthal Moravian Church 21
Frederiksted 41,60-69
gingerbread 60,70i
map 61
tour for cruise ship passengers 13
walking tour 66
Frederiksted Pier 64-65
French Ruins, Estate Judith's Fancy 97
Frenchman Hill (also Mt. Washington in Northside A Quarter) 74
Friedensfeld Moravian Church 88
Frogfishes 65
Furnel Point (Stewart Point, Wells Point) 117
Furnel Ridge 116;117 map;118
Fustick (San Bartolomé) 154

Gallows Bay view 24;137
Gar (Needlefish) 139
Garden Eel 65
Genip Tree 64
Glynn Farms Road 102
Goat Peak (Sugarloaf Hill) 153
Golden Shower Tree 20
gongolo 76
Goose (seaplane) 21
Government House 19,20i
grapes 54
Grapetree Bay Hill Road 159
Grapetree Bay Hotel 160
Grapetree Bay, view 162
Grapetree Beach Hotel 162
Grapetree Point, view 162
Grass Point 162
Grass, Manatee 147
Shoal 147
Turtle 147
Gray Kingbird 41i
Great Blue Heron 166
Great Egret 166
Great Gut 74
Great Pond area 165-167,168i; views 164,168,177
Great Pond 164,165,166
Great Pond berm 166,169
Great Pond Nature Trail 166-167,169
greathouse--see Estate
Greatpond Hill 177; map 176

Greatpond Hill Road 176-177
Green Cay Marina 145
Green Cay National Wildlife Refuge 18,145
Green Cay rookery 145
Green Sea Turtle 57
Green-backed Heron 85,166
Gregre Tree 102,103,132,175,177
Griggs Home for the Aged 29
Gròtto Beach 144
guide books 17,40
gulls 146
Gumbo-limbo (Naked Tourist Tree, Turpentine Tree) 29,44;47 description;134,179
gut = creek, streambed

Halfbeak 139
Halfpenny Bay views 23,138,171;172
Ham Bluff 80; view 124
Ham Peak 116
Hamilton Way 97
Hamilton, Alexander 25
Hawksbill Turtle 57
Heathen Bay 148
Hemer Peninsula 97; views 128
herons 146
Hess Channel, view 23,139
Hess Oil Refinery 52-53; views 23,139
  main gate 52
hike, Annaly Bay 122-123
  Buck Island 179
  Butler Bay Gut 75-78
  Caledonia Gut 79,125
  Columbus' Landing beach 106
  Columbus' Landing mangrove 105
  Creque rain forest 91
  East End beaches 156-157
  East Point 157
  French Pond 173-174
  Furnel Ridge 116-119
  Great Pond Nature Trail 166-167,169
  Ham Bluff lighthouse 80
  Manchenil Bay 171
  Mount Eagle forest 112-113
  Mt. Stewart Gut 83
  Mount Washington 74
  northeast coast 148,149,150,151
  St. George Gut 84
  southeastern coast 172-173
history 9-12

Hodge Bay 150
Hogplum Tree 86,89
Holy Cross Episcopal Church 86
Hope Road (Route 68) 52
Horse Peak 124
horseback riding 73
Horseradish Tree 54
Horsetail Tree (Horsetail Casuarina, Australian-pine) 69
horticultural site: Anna's Hope 25
horticultural site: Buccaneer Hotel 143
  Cruzan Gardens Nursery 87
  Cruzan Motors 99
  Grange Hill Nursery 25
  St. George Botanical Garden 35
  Sprat Hall 73
  U.S.D.A. Experimental Station 32
  University of the Virgin Islands 32
  Virgin Islands Agricultural Station 33
hospital (clinic), Frederiksted 64
  (old), Peter's Farm 21
  St. Croix main 27,99
Hughes Hill 153
Hughes Point 161

Iguana Lizard 153
India Rubber Tree 28
Indian Almond 34,100
Indian ballcourt 106
Indian Rosewood 100
Indians, Arawak 106
  Carib 106
  pre-Columbian 9,106
  Taino 106
instructions 7
ironwork, St. George Botanical Gardens 36
Isaac Bay 157
Isaac Point 157
Island Center for the Performing Arts 27,99

Jack Bay 157,161
Jack Spaniard 111i
Jacobsberg Ridge 22,139
Jealousy Gut 135
Jefferson Way 97
Jellyfish, Upside-down 140,145
John David Merwin Highway 71

Jojoba 54
Judith Fancy Formation 38,85

Kallaloo Crab (Land Crab, White Crab) 103,105,178i
Kapok Tree (Ceiba, Silk Cotton Tree) 67,69,72;89 description;91,92
Kenjo Drive 177,178
Kid Peak (East Point, Point Udall) 157
King Cross Street, Frederiksted 68
King Street (Route 631) Frederiksted 66
King Street, Christiansted 19-21
Kingshill Marl 50
Knight Bay (Cistern Bay) 152; view 159
Knights of Malta 9
Krause Lagoon 55

Lablab 54
Labor Act of 1849 11
lagoon--see mangrove
Lagoon Street, Frederiksted 68
Lamb Bay (Bluegut Bay) 155
Lamb Peak 155
land snail 173
Landmarks Society, St. Croix 40
 Estate Clairmont Park 108
Lang Peak, view 167
Laughing Gull 64
LaVallee Road 131
LaVallee village 109,131
LEAP, Life and Environmental Arts Project 82
Least Tern 58,60,166
Leatherback Sea Turtle 40,59i,142
Levine, Rachael Faucett (monument) 25
library, Christiansted 20
 Frederiksted 63
lighthouse 80
Lignumvitae Tree, Common 19,29,45;47 description;52,64,146,148,173
Lily, West Indian Spider 69,123
Limpricht Park 20
Little Blue Heron 62,64
Little Princesse Hill Road 93
Lizard, Iguana 153
 St. Croix Ground 18,145
 tree 135i
Loblolly = Water Mampoo Tree
Locust, West Indian 112,130
Long Reef, Christiansted 23,95,138; view 24
Louise Augusta, Fort 141
Lowry Hill Road (Route 62) 167-169
Lutheran Church, Christiansted 19

Madsen Point 164
Magnificent Frigatebird 64,166,179
Mahoe, Beach 100
Mahogany Road (Route 76) 81-86
Mahogany Shop 82
Mahogany Tree, Honduran 44,134
 hybrid 29,44,45,49
Mahogany Tree, Pacific 43,44,45
 West Indian 19,21,44,45,66,67,68, 69; 81 description; 102,109, 123,135
Mahogany-Midland Road 72-88
Malta Baths 78i
Maltese Ruins, Estate Judith's Fancy 97
Mammee Apple 86,89,112
Mampoo Tree, Water (Loblolly) 161
Manatee Grass 147
Manchenil Bay 171; views 23,138,171
Manchineel Tree 56,109,149,150,156, 163,167,169,179
 poison 56
Mango Tree 33,54,67i,83,86,88,91
Mangrove Tree, Black 39; 48 description; 49i,53,55,56, 103,105,140,145,146,149,165i, 167,174,179
Mangrove Tree, Red 39;48 description; 49i,53,55,56,103,104i,105,140i, 145,146,149,165,167,174
Mangrove Tree, White 39; 48 description; 53,55,56,63, 103,105,140,145,146,149,165, 167,169,174,179
mangrove, Altona Lagoon 139-140,145; views 24,138,141,142,143
 Alucroix Channel 48-49
 Buck Island 179
 Coakley Bay Saltpond 149
 Fairplain Creek 55
 French Pond 53,174
 Great Pond 165,167
 Krause Lagoon 55
 Long Point 39
 Manning Bay Lagoon 55,56
 Salt River Estuary, Columbus' Landing 105
 Salt River Estuary, Sugar Bay 103

mangrove, Salt River Estuary, Triton Bay 98
Southgate Pond 146
Southgate Saltpond 146
Manioc (Cassava) 54
Manjack, White 148; 150 description
Manning Bay Lagoon 55,56
map, Oxholm 40
maps 7
maps, U.S.G.S. topographic 18
Maria Hill 167
Marina Road 145
Market Street, Frederiksted 67
market, Christiansted 19
Frederiksted 66i,67
Marley, Robert 35
Maroon Hole 80
marsh, Fredensborg Pond 30
Prosperity 72
Rust-op-twist 109
Martel Bay 143
Martin Marietta entrance 29,50
Meggie 161
Melvin Evans Highway 46
Mermaid Beach 143
Merwin, John David 71
Mespel Tree = Sapodilla Tree
Michael Hill 107
Midland Road (Route 72) 86-88
Midland-Mahogany Road 72-88
mill--see Estate
mill openings 120
millipede 76
black 77
brick red 77
brown with yellow-green bars 77
Milord Point, view 168
Miss Bea Road 147
Molly (fish) 83,85
Mon Bijou Road 131
mongoose 18,113
Monkey-pistol Tree (Sandbox Tree) 89 description; 90,92
Monks Baths 78
Morne Rond 156
moth, Black Witch 76
Mt. Eagle 33,114
Mt. Fancy 164
Mt. Fancy Saltpond 162 view;164
Mt. Stewart Gut 82,83i
Mt. Tallin 152
Mt. Washington (Eastend A Quarter) 176
Mt. Washington (Northside A Quarter, also Frenchman Hill) 74
mullets 139
mural, "Tribute to Robert Nesta Marley" 35
museum, Fort Frederik 66
Steeple Building 17
Whim Plantation 40

National Historic Landmark, Columbus' Landing 105
National Historic Site, Christiansted 17-18
National Monument, Buck Island 179
National Wildlife Refuge, Green Cay 18,145
Sandy Point 58-60
Nature Conservancy, The 97,98
Needlefish (Gar) 139
Night-blooming Cereus 62
North Road (Route 75) 93-101
North Shore Road (Route 80) 101-113
North Side Road (Route 75) 93
north side--also see "northside" 93
Northeast Shore Drive 137
Northside Road (Route 63) 71
northside--also see "north side" 71
Nugent Peak 168
nursery, Cruzan Gardens 87
Grange Hill 25

Old Man's Beard 19
Old-woman's-tongue = Tibet Tree
Orangegrove Road (Route 765) 83
orchid, Chocolate Hole 60,179
Chocolate Hole 60
Christmas 80,124
Sandy Pt. 60,179
Vanilla 125
Otaheita 109,150
Oxford Road (Route 763) 82
Oxholm Map 40
Oxholm Road 101
Oyster, Tree 104

Painkiller Tree 54
Palm, Alexander 100
Caribee Royal 33,100
Christmas = Manila Palm
coconut 94i,95,97,102,109,110,113,148,162
Date 100

Palm, Desert Fan 100
Dwarf Royal = Manila Palm
fan 100
Fiji Fan 100
Manila 100
Puerto Rican Royal 69,83,86,100
royal 99
Solitaire = Alexander Palm
South American Royal = Caribee Royal Palm
Tyre 117
Puerto Rican = Puerto Rican Royal Palm
Palmetto, Puerto Rican 100
Papaya Tree 54
passion fruit 54
Pasture Fiddlewood 69
patch reef, Boiler Bay 155
Cottongarden Point 153
Great Pond Bay 164
Knight Bay 152,160
Robin Bay 164
Peak Seven, views 138,168
Pearly-eyed Thrasher 77,91
Pelican Cove 97
Pencil Tree 34,88
Pentheny Peak 162
Pepper Tree Road (Route 79) 98
Pigeon Peas 54
pineapple 54
Pink-cedar Tree = White-cedar Tree
Pitch-apple 101
plovers 146
Point Cudejarre 157
Point Udall (East Point, Kid Peak) 157
police station, Christiansted 19
Frederiksted 40
Golden Grove 33,51
LaReine 28,101
porpoise 160
post office, Christiansted 17
Frederiksted 41,61
Golden Rock 21
Kingshill 101
pre-Columbian artifacts 17,69
pre-Columbian Indians 106
Primavera 80
Prince Street, Frederiksted 67
prison, old 21
prison, territorial 51,55
Prosperity Marsh 72
Protestant Cay 17,18i; 24 view

Prune Bay 147
Pull Point 148; views 24,138
Punnett Bay 144

Quassia Hill, view 23
Queen Mary Highway 24
Queen Street, Frederiksted 67-68
Questa Verde condominiums 22

radio station WSTX view 24;141
rain forest, Creque Dam Road 72,88-92
Mahogany Road 72,81-85
Raintree (Samán) 71,72,79,82,86,92, 135
Ramirez Road 29
Rastafarian 35
Recovery Hill 137-139
Red Bay (Rod Bay) 163
Reef Condominium 151
Reef Convenience Store 152
reef, Pleistocene fossil 48
rental cars 8
River Road (Route 69) 86,114,133-135
Robin Bay 164; view 162
Rocky Hill 161
Rod Bay (Red Bay) view 162;163,164
Roger Gut 80
Romney Hill 159
Romney Point 152
rookery, Buck Island 179
Green Cay 145
Rosewood, Indian 100
Round Reef, view 24
Route : j.=junction
Route 58 (Creque Dam Road) 72,88-92
j. with Route 78 122
Route 60 (Southshore Road) 159-167
Route 60 (Tipperary Road) 175-178
(eastern) j. with Route 82 152
(western) j. with Route 82 146
j. with Route 624 167
Route 62 (Lowry Hill Road) 167-169
Route 62 (Southside Road) 167-174
j. with Route 68 54
j. with Route 82 142
Route 63 (West End Road) 71-80
j. with Route 58 88
j. with Route 66 52
j. with Route 70 40
j. with Route 76 81
j. with Route 78 125
j. with Route 631 69

Route 631, King Street, Frederiksted 66
Route 64 (Airport Road) 54-57
(eastern) j. with Route 70 32,51
(western) j. with Route 70 36,51
Route 66 (Four-lane Highway) 46-51
(eastern) j. with Route 64 55
(western) j. with Route 64 57
j. with Route 70 27
Route 68 (Hope Road) 52-54
j. with Route 62 174
j. with Route 66 47
Route 69 (River Road) 86,114,133-135
j. with Route 70 33
j. with Route 78 131
j. with Route 80 113
Route 69, Route 70 to Route 66 W. 34
Route 70 (Centerline Road) 24-41
j. with Route 62 174
j. with Route 63 71
(eastern) j. with Route 64 54
(western) j. with Route 64 57
j. with Route 69 135
j. with Route 79 99
j. with Route 81 96
j. with Route 83 173
j. with Route 661 58
j. with Route 663 133
j. with Route 708 43
Route 72 (Midland Road) 86,88
j. with Route 69 135
j. with Route 73 132
j. with Route 75 99
Route 73 (Canaan Ridge Road) 129,131-133
j. with Glynn Farms Road 102
j. with Route 70 28
j. with Route 72 88
j. with Route 78 129
j. with Route 80 109
j. with Route 707 31
Route 74 95
j. with Route 75 95
j. with Route 79 98
Route 75 (North Road) 93-101
j. with Route 70 28
j. with Route 753 21
Route 76 (Mahogany Road) 81-86
j. with Route 63 72
j. with Route 69 135
Route 78 (Scenic Drive E.) 127-129
Route 78 (Scenic Drive W.) 115,122-125,129-131
Route 78 (Scenic Drive W.),
j. with Route 58 92
j. with Route 69 114,133
j. with Route 73 132
j. with Route 765 84
Route 79 (Pepper Tree Road) 98
j. with Route 70 27
j. with Route 75 98
Route 80 (North Shore Road) 101-113
j. with Route 69 133
j. with Route 73 131
j. with Route 75 99
Route 81 43,95
j. with Route 68 53
Route 81, j. with Route 70 26
j. with Route 74 95
Route 82 (East End Road) 137-158
Route 83 173
j. with Route 62 173
j. with Route 70 24
j. with Route 75 93
Route 85 (Springgut Road) 170
j. with Route 62 170
j. with Route 82 139
Route 624 167
j. with Route 60 167
Route 661 40,58
j. with Route 66 52
j. with Route 70 40
Route 663 29,133
j. with Route 66 50
j. with Route 70 28
Route 682 25
j. with Route 62 174
j. with Route 68 53
Route 701 39
j. with Route 70 40
Route 702 71
Route 703, j. with Route 66 52
j. with Route 70 39
Route 705 35
Route 707 31
j. with Route 70 30
j. with Route 73 132,133
j. with Route 75 101
j. with Route 663 133
Route 708 43
j. with Route 70 26
Route 709 28
j. with Route 75 101
Route 751 97
Route 753 22

Route 763 (Oxford Road) 82
j. with Route 765 84
Route 765 (Orangegrove Road) 83
j. with Route 58 92
j. with Route 763 82
Route 811 43
Royal Palm, Caribee (South American Royal Palm) 33,100
Royal Palm, Puerto Rican 69,83,86,100
royal palms 99
ruins--see Estate
rum distillery, Brugal 72
Cruzan Rum 57
Ruth Island 48; views 23,139
Saba, views 153,157
Sahara dust 114
salt flats, Great Pond 166
Salt River Bay 97
Salt River Estuary, Sugar Bay 103
Salt River Estuary, Triton Bay 98
Salt River Submarine Canyon 128
saltpond, Altona Lagoon 144
Buck Island 179
Coakley Bay 148-149
French Pond 174
Great Pond 165-167
Long Point 39
Mt. Fancy 164; view 162
Southgate 146
Westend 40,58,63
Saltriver Mountain 128
Saltriver Point 98
Samán = Raintree
San Bartolomé (Fustick) 154
Sandbox Tree (Monkey-pistol) 89 description; 90,92
Sandy Point 40,59,62
Sandy Point National Wildlife Refuge 58-60
Santa-maría 101
Sapodilla (Chicle, Mespel Tree) 54,58
Sausage Tree 27,53,100
Scale House, Christiansted 18
Frederiksted 65
Scaly-naped Pigeon 150
Scenic Drive E. (Route 78) 127-129
Scenic Drive W. (Route 78) 115,122-125,129-131
Scenic Drive, for cruise ship passengers 13
schoolhouse, Diamond 36
Frederiksted 68
Green Cay (Easthill School) 148
Kingshill 29
LaGrande Princesse 96
Mount Victory 91
Peter's Rest 26
Schoolmaster 140
Scotch Bank, views 24,138,142
Scratch Peak 123
Screwpine, Common 100
SCUBA--see dive, diving, snorkel, snorkeling
sea fans, Coakley Bay 149
Sea Hibiscus = Beach Mahoe
Sea Urchin, Rock-boring 58,79,154
sea whips, Coakley Bay 149
Malta Baths 79
Seagrape Tree 60,141,146,150,162,172
seagrass bed, Boiler Bay 155
Butler Bay 74,75
Fareham Bay 173
Great Pond Bay 164
Halfpenny Bay 173
Jack Bay 157
Manchenil Bay 171
Robin Bay 164
Rod Bay 163
Romney Point 152
Salt River Bay 98
Smuggler's Cove 152
Southgate 147
Spring Bay 173
Turner Hole 162
Seahorses 65
seaplane terminal 21
seeds, drift 156
Semi-palmated Plover 166
Seven, Peak, view 138
Shek-shek = Tibet Tree
shelling, East End Bay 156
Sandy Point 60,62
Sheltered Workshop 29,30i
shipwreck, Cumulus 98
Shoal Grass 147
shopping center, Golden Rock 21,93
Sunny Isle 27,46,53
Villa LaReine 28,101
Shortleaf Fig 62,78,149
Shoy Point 141,143
Shoy's Beach 142
Shoy's Washer view 141;143
Silk Cotton Tree (Ceiba, Kapok Tree) 67,69,72; description 89; 91,92

Silver Tabebuia Tree 99,134
Smuggler's Cove 152,159
snail, land 173
Snake Plant 34
snappers 139
snorkel, Altona Lagoon 140
Bluegut Bay (Lamb Bay) 155
Boiler Bay (Brown Bay, Colorado River, Whale Bay) 155
Buck Island 179
Butler Bay 74
Cane Bay 110
Coakley Bay 149
Cottongarden Point 153
Fairleigh Dickinson Territorial Park 153,154,155,156,157
Fort Louise Augusta 141
Frederiksted Pier 65
Frederiksted public beach 69
Grapetree Bay 160
Grass Point 163
Great Pond Bay 164
Ham Bay 79
Ham Bluff 80
Isaac Bay 157
Jack Bay 157
Long Reef 95,97
Malta Baths 79
(near) Manchenil Bay 172
Meggie (Grapetree Bay) 161
Pelican Cove 97
Prune Bay 148
Robin Bay 164
Rod Bay 163
Romney Point 152
Salt River Bay 98
Smuggler's Cove 152
Southgate Beach 147
Tague Point 151
Turner Hole 162
Yellowcliff Bay 151
snorkeling tours 17,64,143,151
Solitude Bay 150
Solitude Country Store 150
Solitude Valley Road 150
Sorghum 54
soursop 67i,102
South Side Road (Route 60) 159
Southgate Beach 145
Southgate Saltpond 143 view; 145,146
Southgate-Great Pond Road 175
Southshore Road (Route 60) 159-167
Southshore Road (Route 62) 167
Southside Road (Route 62) 167-174
Southwest Point 59
Spanish Moss 22,90i,91
Spider Lily, West Indian 69,123
Spider, Golden Silk Orb Weaver 76,85, 118,119i
Spirula, Common 156
sponges, Coakley Bay 149
Frederiksted Pier 65
Malta Baths 79
Sprat Hall Plantation 73,89
Sprat Hall Valley Road 88
Spring Bay 169,172; view 171
Springgut Road (Route 85) 170
St. Croix Agave 164,171,172i
St. George Gut 84
St. George Hill 37; view 122
St. John's Anglican Church 21
St. Joseph's Chapel 35
St. Patrick's Church 67
St. Paul's Episcopal Church 68
Stable, Jill's Equestrian 73
steel (band) orchestra 63
Steeple Building 17,19
Stewart Bay (Wills Bay) 118,123
Stewart Point (Furnel Point, Wells Point) 117
Stony Ground Beach 63
Strand Street, Frederiksted 62-66
Strangler Fig 75,76i,84,86,134
Strangler Fig (Shortleaf) 62,78,149
Strangler Fig--also see Pitch-apple, Santa-maría
Sugar Bay 103
sugar cane 54,102
sugar factory, Bethlehem Old Works 87,101
Estate St. John 98
--also see Estate
Sugar-apple 67i
Sugarloaf Hill (Goat Peak) 153
Sunny Isle Intersection 27,46,53
supermarket, Golden Rock 21,93
Sunny Isle 27
Villa LaReine 28,101
surfing, board, Salt River Bay 98
surfing, body, Grapetree Bay Beach 160
Meggie 161
Punnett Bay 142
Shoy's Beach 142
Sweet-lime 46,84

Tabebuia, Silver Tabebuia Tree 99,134
Tague Bay 151-152; views 151,152,159
Tague Point 151
Taino Indians 106
Tamarind Reef Road 146
Tamarind Tree 68,69,89,167
Tantan Tree 47,54
taxis 8
Teak Tree 6i,42i,44,97,120
Termite, Long-nosed 15
terns 146
Thunder Hole (Turner Hole) 162
Thunderhole Hill 162
Tibet Tree 71,86
tidepools, Annaly Bay 119,123
  Annaly Bay Notch 119,123
  Chenay Bay 148
  Gentle Winds 106
  Grass Point 163
  Ham Bluff 80
  LaVallee 110
  Malta Baths 79
  Northside 79
  Wills Bay 118
Tilapia 54
Tipperary Road (Route 60) 175-178; map 176
Toad, Giant (Crapo) 42
Touch-me-not 149,179
tourist bureau, V.I., Christiansted 18
  Frederiksted 66
Tourist Tree = Gumbo-limbo Tree
Triton Bay 98; view 97
Turk's Cap Cactus 155i,157,179
Turner Hole (Thunder Hole) 162,163i; views 161,162
Turpentine Tree = Gumbo-limbo Tree
Turtle Grass 147
Turtle, Green Sea 57
  Hawksbill Sea 57
  Leatherback Sea 40,59i,142
Tyre Palm 117

University of the Virgin Islands Extension Campus 32

Vagthus Point (Watchho) 171;views 23,139,171
Vallee View Road 130
Vanilla Orchid 125
Veterans' Road 71
Victoria House 64
view, Altona Lagoon, Christiansted Harbor 143
  (from) Bodkin mill 119
  both north and south shores at mid-island 96
  Buck Island 150,151
  Buck Island and Tague Bay 151,152
  Christiansted Harbor 23,94,137-138,141
  Columbus' Landing 128
  (from) Cotton Valley mill 150
  (from) Fort Louise Augusta 141
  Fountain Valley 134
  Frederiksted 73,74,82
  Frederiksted and western shore 71
  Great Pond area 164,168,177
  (from) Greatpond Hill 177
  Ham Bluff, Puerto Rico, northern V.I. 124
  industrial complex 29,127
  Knight Bay 159
  (from) Maria Hill 167
  Maroon Ridge 108,116,130
  (from) Michael Hill 107
  middle of island 23
  Mount Eagle 110
  (from) Mt. Washington (Northside A Quarter) 74
  north- and southeastern coasts 167
  north- and southwestern coasts 119,120,121,129
  northeastern coast 150,175,177
  northern and southern coasts 23
  northern coast 107,108
  northwestern coast 80,115,116,123,124
  Puerto Rico 116
  Saba 153,157
  (from) St. George Hill 37
  Salt River Bay 127
  Salt River Bay and northern coast 127,128
  (from) Seven Hills 176
  southeastern coast 162,171,177
  southwestern coast 37,122,130
  Tague Bay 159
  Triton Bay 97
  Virgin Islands, northern 24,114,116,133,138,153,157
  western coast 74
Vincent Mason Coral Resort 63
vine, Cissus 79,112,135

Virgin Islands, northern, view 24, 114,116,124,133,138,153,157
Virgin Islands, University of the 32
visitors' bureau, V.I., Christiansted 18
Frederiksted 66
von Scholten, Governor Peter 11,26,66

walking tour, Christiansted 18
Frederiksted 66
WAPA, view 23,138
warblers 103,145,146
wasp 111
Watchho (Vagthus Point) 171; views 23,139,171
Water Mampoo Tree (Loblolly) 161
watersports 17,151
watersports--also see diving, fishing, snorkeling
Wells Bay 117,122,123
Wells Point (Furnel Point, Stewart Point) 117
West End Road (Route 63) 71-80
West Indian Cedar (Cigar Box Cedar) 46,80
West Indies Laboratory 152
West Shore Tour, for cruise ship passengers 13
Westend Saltpond 40,58,63
Whale Point 155
Wharf Square 17
whim 57,62
Whim Plantation Museum (restored animal mill, restored greathouse, museum, sugar factory, restored windmill) 39,40i
Whistle Beach 144
White Fig = Shortleaf Fig
White Horse 128
White Manjack 148,150 description
White's Bay 58
White-band Disease 160
White-cedar Tree (Pink-cedar Tree) 50,53,86,134
White-cheeked Pintail 146,174
White-crowned Pigeon 60,146,166,174
White-tailed Tropicbird 53,171
Whitehorse 98
Wild Fig = Pitch-apple
Wild Tanier 117,118i
Wills Bay (Stewart Bay) 118,122,123
Wilson's Plover 166
Windward Bay (East End Bay) 156
Windwood Estates Loop 128
Wismenog Point 148
wood crafts 82
Woodcottage Bay 163

Yellow Flamboyant 64
Yellow Warbler 60,166
Yellow-bellied Sapsucker 103,132
Yellow-crowned Night Heron 85
Yellowcliff Bay 151

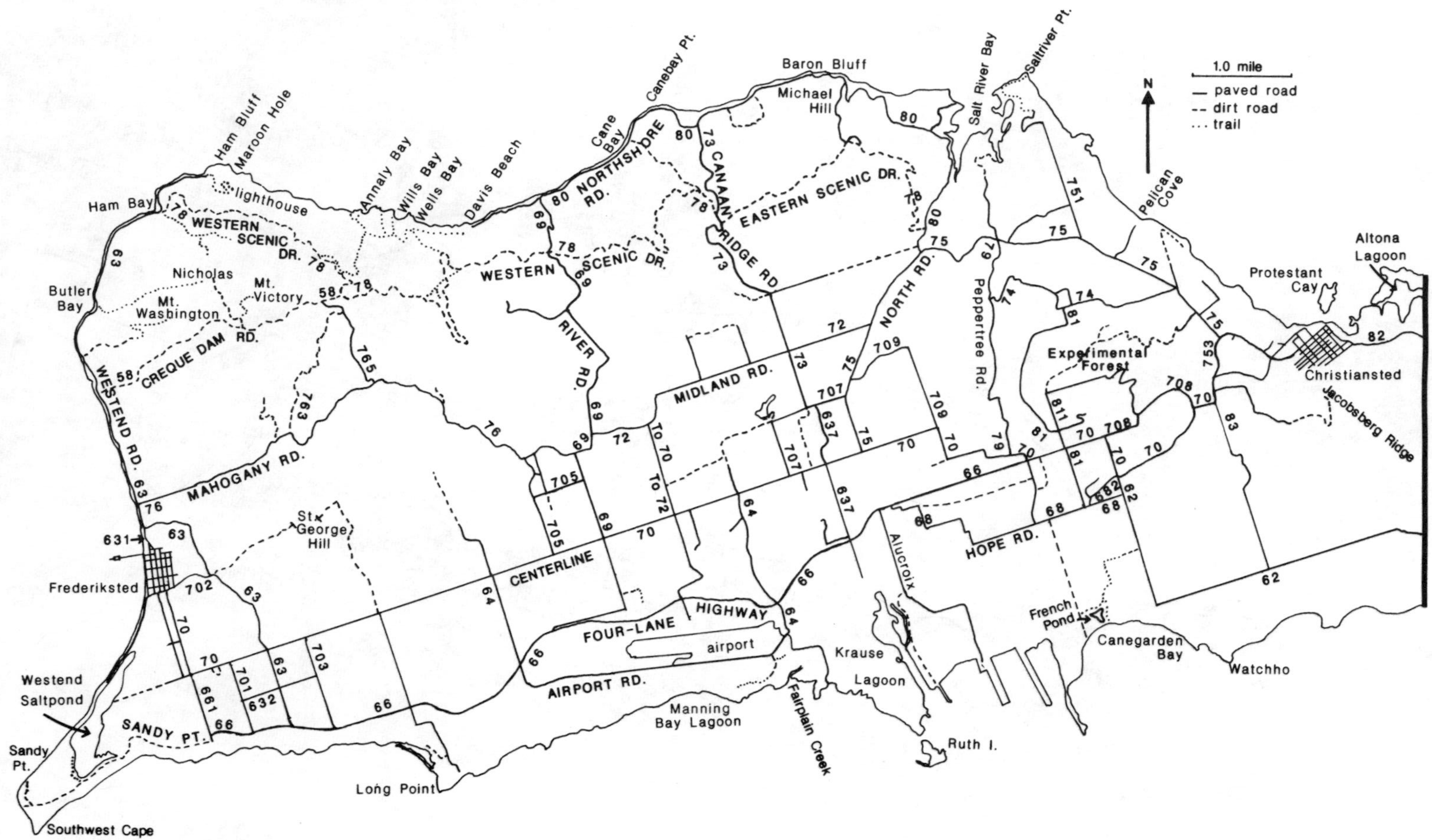
1.0 mile
— paved road
-- dirt road
... trail
N
Baron Bluff
Michael Hill
Canebay Pt.
Cane Bay
Salt River Bay
Saltriver Pt.
Pelican Cove
Altona Lagoon
Protestant Cay
Christiansted
Jacobsberg Ridge
Ham Bluff
Maroon Hole
Ham Bay
lighthouse
Annaly Bay
Wills Bay
Wells Bay
Davis Beach
WESTERN SCENIC DR.
EASTERN SCENIC DR.
NORTHSHORE RD.
CANAAN RIDGE RD
NORTH RD.
Peppertree Rd.
Experimental Forest
Butler Bay
Nicholas
Mt. Washington
Mt. Victory
CREQUE DAM RD.
WESTERN SCENIC DR.
RIVER RD.
MIDLAND RD.
WESTEND RD.
MAHOGANY RD.
St George Hill
To 70
To 72
CENTERLINE
HOPE RD.
Alucroix
Frederiksted
FOUR-LANE HIGHWAY
airport
AIRPORT RD.
French Pond
Canegarden Bay
Watchho
Krause Lagoon
Ruth I.
Fairplain Creek
Manning Bay Lagoon
Westend Saltpond
SANDY PT.
Sandy Pt.
Long Point
Southwest Cape

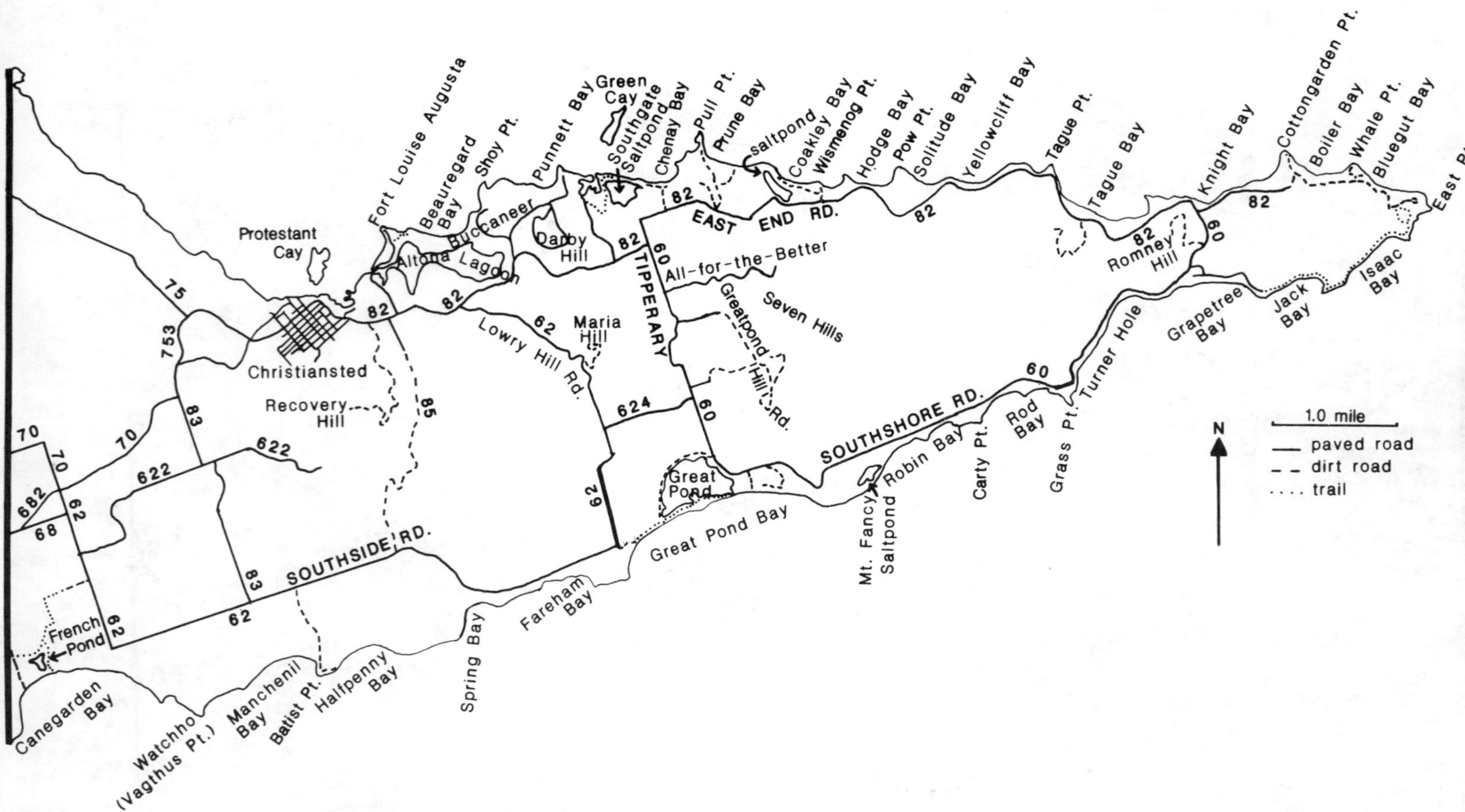
Fort Louise Augusta
Beauregard Bay
Shoy Pt.
Punnett Bay
Green Cay
Southgate Saltpond
Chenay Bay
Pull Pt.
Prune Bay
Saltpond
Coakley Bay
Wismenog Pt.
Hodge Bay
Pow Pt.
Solitude Bay
Yellowcliff Bay
Tague Pt.
Tague Bay
Knight Bay
Cottongarden Pt.
Boiler Bay
Whale Pt.
Bluegut Bay
East Pt.
Protestant Cay
Buccaneer
Altona Lagoon
Darby Hill
Christiansted
Recovery Hill
Maria Hill
Lowry Hill Rd.
TIPPERARY
EAST END RD.
All-for-the-Better
Seven Hills
Greatpond Hill Rd.
Romney Hill
Isaac Bay
Jack Bay
Grapetree Bay
Turner Hole
Grass Pt.
Rod Bay
Carty Pt.
Robin Bay
SOUTHSHORE RD.
Mt. Fancy Saltpond
Great Pond
Great Pond Bay
Fareham Bay
Spring Bay
Halfpenny Bay
Batist Pt.
Manchenil Bay
Watchho (Vagthus Pt.)
Canegarden Bay
French Pond
SOUTHSIDE RD.
1.0 mile
N
paved road
dirt road
trail
75
753
70
682
68
62
83
622
82
85
624
60